
Semiconductor Circuits: Theory, Design and Experiment

Semiconductor Circuits: Theory, Design and Experiment

by

J. R. ABRAHAMS, B.SC.(ENG.), M.SC., M.I.E.E.

and

G. J. PRIDHAM, B.SC.(ENG.), M.I.E.E., M.I.E.R.E.

Senior Lecturers, Enfield College of Technology

1966

PERGAMON PRESS

OXFORD · LONDON · EDINBURGH · NEW YORK

TORONTO · SYDNEY · PARIS · BRAUNSCHWEIG

Pergamon Press Ltd., Headington Hill Hall, Oxford
4 & 5 Fitzroy Square, London W.1
Pergamon Press (Scotland) Ltd., 2 & 3 Teviot Place, Edinburgh 1
Pergamon Press Inc., Maxwell House, Fairview Park, Elmsford, New York 10523
Pergamon of Canada Ltd., 207 Queen's Quay West, Toronto 1
Pergamon Press (Aust.) Pty. Ltd., 19a Boundary Street, Rushcutters Bay, N.S.W. 2011, Australia
Pergamon Press S.A.R.L., 24 rue des Écoles, Paris 5^e
Vieweg & Sohn GmbH, Burgplatz 1, Braunschweig

First edition 1966

Reprinted 1970

Library of Congress Catalog Card No. 65–29054

Printed in Great Britain by Bell and Bain Ltd., Glasgow, and reprinted lithographically by A. Wheaton & Co., Exeter

08 011651 5 (flexicover)
08 011652 3 (hard cover)

Contents

Introduction

THIS book is the result of a number of years of experience teaching engineering undergraduates and senior technicians. Both of the authors have been concerned with the design of transistor circuits from the days of the early point contact devices.

Initially, we hesitated before adding to the extensive literature on semiconductor topics until we were convinced that the book could fill a significant gap in that literature. It has been written specifically for the student, whether he is on a full- or a part-time course. A major aim is to give each student, of electronic engineering, confidence in his ability to design and modify circuits involving transistors and related semiconductor devices.

Due to space limitations some topics that are covered extensively elsewhere, e.g. smoothing circuits, are treated superficially in these pages. Particular aspects, e.g. load line construction, are covered in detail in the companion Volume of Worked Examples.

This textbook is divided into three clearly defined parts. The four chapters of Part I are concerned with the physical theory of semiconductors; the production of *pn* junctions; the characteristics and equivalent circuits of transistors. The treatment of physical theory is necessarily brief, but has been included since engineering students require some form of model on which to build.

Part II forms the major portion and is made up of seven chapters. These chapters have been written at a practical level, and include a number of complete circuit designs. Chapters 10 and 11, in particular, deal with aspects that have been neglected in previous textbooks on semiconductors. The majority of these circuits have been designed for *pnp* transistors. However, a simple change of supply polarity and transistor symbol gives the corresponding circuit for *npn* transistors.

Some twenty demonstrations or experiments are given in Part III. The staff at Enfield College have devoted much effort to the development of worth-while laboratory work, and these three chapters present some of the results of this endeavour. This could prove useful if the reader is concerned with the organization of semiconductor courses.

A few questions are given at the end of each chapter in Parts I and II. Some of these questions have been designed to emphasize aspects of the text, and others have been taken from past examination papers in various subjects. One or two questions in some chapters are intended to lead the student on to further reading. The sources of these questions are indicated at the end of the question by the following abbreviations:

(B.Sc.), London University, external B.Sc.(Eng.), Part III. Telecommunications or Electronics.

(I.E.E.), Institution of Electrical Engineers, Part III. Applied Electronics or Electronic Engineering.

(H.N.D.) or (H.N.C.), Higher National Diploma or Certificate of Enfield College.

The University or Institution is in no way responsible for the numerical answers given in Appendix A.

Acknowledgements

THE material of this book has been used with a large number of students of electronic engineering in various courses at Enfield. We owe a debt to these students for their patience with our developing efforts in this subject, for many worth-while queries, and for some down-to-earth advice.

Over the past seven years a number of specialists in the semiconductor field have given lectures in our department. Some of the ideas from these lectures have had considerable influence on our writing.

Mr. N. Hiller, who is editor of this series, has been very helpful at various stages in the preparation of this manuscript.

We are grateful for the permission of London University and the Institution of Electrical Engineers to use certain questions taken from past examination papers of those two bodies.

Four of the figures in Chapter 10 are reproduced from the paper " Generation of Power in Satellites ", by H. J. H. Sketch, which was published in the *I.E.E. Journal*, Vol. 8, 1962, p. 148.

Finally, we wish to acknowledge the permission of the Governors of Enfield College (through the Head of the Electrical Engineering Department) for this book to be published.

J. R. A.

G. J. P.

PART I

Theory of Semiconductors

CHAPTER 1

Basic Physical Theory

THE understanding of semiconductor diodes, transistors and the many other solid state devices that have been developed in the past decade involves a deeper knowledge of the physical structure matter than is required with the well-established valve circuits. In this chapter, the underlying physical principles will be developed and, in the next, applied to semiconductor devices.

1.1. The Atom

The atom of any element is the smallest particle of that element capable of taking part in a chemical reaction. It may be regarded as being composed of electrons moving in circular or elliptical orbits about a relatively heavy nucleus of protons and neutrons as shown in Fig. 1.1. This model of the atom was proposed by Bohr in 1913 and although it has been displaced by later models for many purposes it is a convenient representation to show the action of semiconductor devices.

Electrons have a mass of 9×10^{-31} kg and carry a charge of $1{\cdot}6 \times 10^{-19}$ C, while protons have an equal positive charge but a mass 1838 times as great. Neutrons are about the same mass as protons but carry no charge.

All atoms are about the same size, approximately 10^{-10} m in diameter, while the nucleus is about 10^{-15} m in diameter. The nuclei of the heavier elements provide a greater attractive force on the orbital electrons and constrict the electrons within approximately the same atomic volume. The atomic number of an element is decided by the number of protons in the nucleus while the atomic weight is governed by the number of protons and neutrons.

1.2. Electron Orbits

The position and shape of the electron orbits were derived by Bohr who applied Planck's Quantum Theory to atomic structure. This theory states that the energy of a body can only change by definite units of energy known as quanta and by applying this idea to the production of light by gas discharge tubes Bohr showed

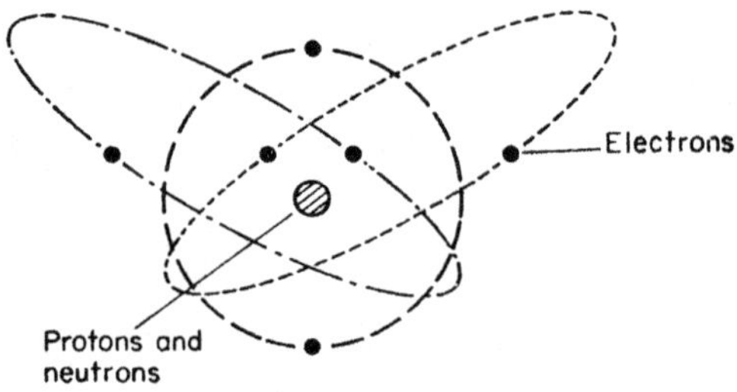

FIG. 1.1. Structure of the atom.

that definite electron energy levels existed. The movement of electrons from a high energy level E_1 to a low energy level E_2 is accompanied by the radiation of energy according to the relation:

$$E_1 - E_2 = h.f., \quad \text{where } h = 6{\cdot}624 \times 10^{-34} \text{ J-sec}$$

i.e. Quanta of Energy = (Planck's Constant) × (frequency of radiation).

This theory gave rise to the Quantum Numbers of electron orbits.

1. Principal Quantum Number (n): For a single atom the electrons are arranged in shells of definite energy level where the electrostatic force of attraction is balanced by the centrifugal force on the electron. The principal quantum number (n) which has

integer values gives the energy levels in these shells in arbitrary units. The value of these units depends on the element since the heavier elements with a greater positive charge on the nucleus give rise to a greater attractive force. That is, $n = 1$ corresponds to the K shell, $n = 2$ the L shell, etc. Thus a shell contains all the electrons with the same principal quantum number.

2. Subsidiary Quantum Number (l): This is a measure of the eccentricity of the electron orbits and has a slight effect on the electron energy. It may have values of $0 \leqslant l \leqslant n-1$ and corresponds to the subshells. That is, $n = 1$, l can only have one value 0, i.e. 1 subshell, $n-2$, l can have values 0 or 1, i.e. 2 subshells. Thus a subshell contains all the electrons with the same principal and subsidiary quantum numbers.

3. The Magnetic Quantum Number (m): This is associated with the rotation of the electrons about the nucleus and has a slight effect on the energy and may have values $-l \leqslant m \leqslant l$,

i.e.
$$l = 0, \quad m = 0,$$
$$l = 1, \quad m = -1, 0 \text{ or } 1,$$
$$l = 2, \quad m = -2, -1, 0, 1 \text{ or } 2.$$

4. The Spin Quantum Number (m_s): This is associated with the rotation of the electron about its axis and may have two values $\pm\frac{1}{2}$.

The application of Pauli's Exclusion Principle which states that no two electrons may have the same four quantum numbers will now give the number of possible electron orbits.

$n = 1$, K shell, $l = 0$, $1s$ subshell, $m = 0$,
$m_s = \pm\frac{1}{2}$, 2 electrons.

$n = 2$, L shell, $l = 0$, $2s$ subshell, $m = 0$,
$m_s = \pm\frac{1}{2}$, 2 electrons.

$l = 1$, $2p$ subshell, $m = -1, 0, 1$,
$m_s = \pm\frac{1}{2}$, 6 electrons.

$n = 3$, M shell, $l = 0$, $3s$ subshell, $m = 0$,
$m_s = \pm\frac{1}{2}$, 2 electrons.

$l = 1$, $3p$ subshell, $m = -1, 0, 1$,
$m_s = \pm\frac{1}{2}$, 6 electrons.

$l = 2$, $3d$ subshell, $m = -2, -1, 0, 1, 2$,
$m_s = \pm\frac{1}{2}$, 10 electrons.

$n = 4$, N shell, $l = 0$, $4s$ subshell, $m = 0$,
$m_s = \pm\frac{1}{2}$, 2 electrons.

$l = 1$, $4p$ subshell, $m = -1, 0, 1$,
$m_s = \pm\frac{1}{2}$, 6 electrons.

$l = 2$, $4d$ subshell, $m = -2, -1, 0, 1, 2$,
$m_s = \pm\frac{1}{2}$, 10 electrons.

$l = 3$, $4f$ subshell, $m = -3, -2, -1, 0, 1, 2, 3$,
$m_s = \pm\frac{1}{2}$, 14 electrons.

This shows that the K shell has two possible electron states, the L shell 8, the M shell 18 and the N shell 32. The number of electron states in any shell is twice the square of the principal quantum number.

1.3. Types of Elements

Elements are classified according to the number of electrons in the outer orbits, the structure being such that there is a maximum of eight active electrons affecting the physical and chemical characteristics.

1. Group 0 elements have eight electrons in their outermost orbits with complete s and p subshells. This is the structure of the inert gases, and neon, with a complete K shell and full s and p subshells within the L shell is an example of this group. Helium, atomic No. 2, is also included in this group since it comprises the complete K shell.

2. Group 1 elements have one electron in the *s* subshell and this group includes the best conductors copper, silver and gold.

3. Group 2 elements have complete *s* subshells.

4. Group 3 elements have three electrons in their outermost orbits and the elements which have complete *s* subshells and one electron in the *p* subshell are important as acceptor impurities added to pure semiconductors. The term acceptor impurity will be explained later in the chapter and such elements used are aluminium, gallium and indium.

5. Group 4 elements have four electrons in their outermost orbits and the ones with a complete *s* subshell and two electrons in the *p* subshell form the basic semiconductor materials. In practice carbon and tin are unsuitable for reasons explained later and the only suitable elements are germanium and silicon.

6. Group 5 elements have five electrons in their outermost orbits and the ones with complete *s* subshells and three electrons in the *p* subshell are important as donor impurities. The elements used are phosphorus, arsenic and antimony and as with group three elements their effect will be explained later in the chapter.

7. Group 6 elements have six electrons in their outermost orbits and are generally insulators.

8. Group 7 elements with seven electrons are one electron short of the inert gas structure.

9. Group 8 elements include those with pronounced individual properties such as the magnetic materials. They have one or two electrons in the outermost orbits with incomplete inner subshells.

1.4. Atomic Binding

The bonds between atoms may be formed in several ways, and a common method is for atoms to borrow electrons from, or share electrons with, other atoms to form the inert gas structure. Semiconductors use the latter method with each atom sharing its four outer electrons with four neighbouring atoms. A two-dimensional sketch of this is shown in Fig. 1.2 and the central atom is seen to have eight electrons under its influence. Two neighbouring atoms and their associated pair of electrons form a covalent bond, and with pure or intrinsic germanium or silicon these bonds must be broken to provide electrons available for conduction.

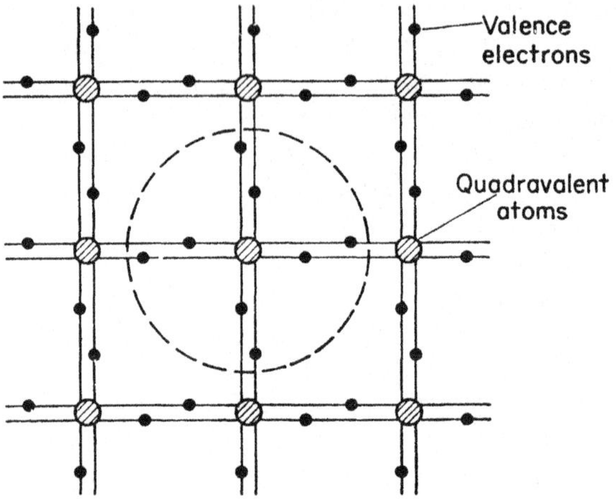

FIG. 1.2. Representation of covalent bonds.

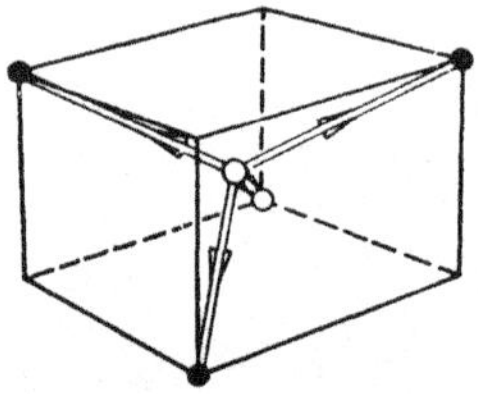

FIG. 1.3. Cubic lattice of semiconductors.

A three-dimensional picture of the atomic structure of a semiconductor crystal shows it to be in the form of a body-centred cubic lattice as shown in Fig. 1.3 and the central atom shares its electrons with the four corner atoms. The complete crystal is made up of many such cubes joined together at their edges as shown in Fig. 1.4 and the corner atom shown will have covalent bonds with four central atoms.

1.5. Electron Energy Levels

The idea of different energy levels may be easily understood by considering the simplest case, which is a single hydrogen atom with unit charge on the nucleus and a single orbital electron. There are definite prescribed orbits and normally the lowest energy orbit is occupied, but if the electron is raised to a higher energy orbit, it will eventually fall back to the stable lowest energy orbit, with the emission of light characteristic of hydrogen.

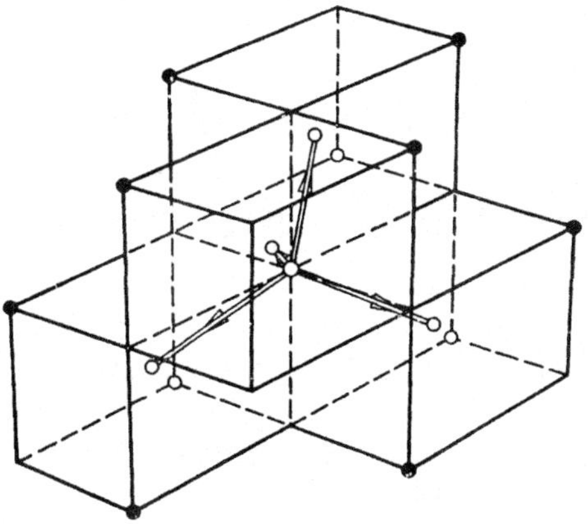

FIG. 1.4. Covalent bonds in three dimensions.

Bohr's basic assumptions are that the angular momentum of an electron in any stable orbit must be an integral number of $h/2\pi$ (where h is Planck's constant) and in such orbits the electrons do not radiate electromagnetic energy. Thus if the moment of inertia of an electron is I and its angular velocity is ω,

$$I\omega = \frac{nh}{2\pi}, \quad \text{where } n \text{ is an integer.}$$

i.e. $mr^2 \cdot \frac{v}{r} = \frac{nh}{2\pi}$, where m is the mass of an electron in kilograms, r is its distance from the nucleus in metres, v is its linear velocity in metres per second.

i.e.

$$mrv = \frac{nh}{2\pi}. \tag{1}$$

Then equating the centrifugal force to the attractive force between electron and nucleus

$$\frac{mv^2}{r} = \frac{e^2}{4\pi\epsilon_0 r^2}, \tag{2}$$

where e is the charge on an electron, ϵ_0 is the permittivity of free space $= 1/(36\pi \times 10^9)$ F/m. This equation is in rationalized mks units. Eliminating v between eqns. (1) and (2) gives

$$r = \frac{n^2h^2\epsilon_0}{\pi e^2 m}. \tag{3}$$

This equation gives radii of possible orbits. The symbol n corresponds to the principal quantum number and in this case the possible orbits are circular. The potential energy in each of these orbits may be found by calculating the work done by an electron moving from infinity to a radius r,

i.e.

$$\text{P.E.} = \int_{\infty}^{r} \frac{e^2 \mathrm{d}r}{4\pi\epsilon_0 r^2} = -\frac{e^2}{4\pi\epsilon_0 r}. \tag{4}$$

The negative sign indicates that work must be done on the electron to remove it from the nucleus.

$$\text{Kinetic energy} = \tfrac{1}{2}mv^2 = \frac{e^2}{8\pi\epsilon_0 r},$$

$$\text{Total energy} = \text{K.E.}+\text{P.E.} = -\frac{e^2}{8\pi\epsilon_0 r} = -\frac{e^4 m}{8\epsilon_0^2 n^2 h^2}. \tag{5}$$

Putting $n = 1, 2$, etc., gives the various electron shells as shown in Fig. 1.5.

Using the values of e, m, ϵ_0 and h previously quoted and substituting $n = 1$ gives

$$r = 0{\cdot}529 \times 10^{-10}\ \text{m}$$
$$= 0{\cdot}529\ \text{Å}$$

and

$$\text{Total energy} = -217 \times 10^{-20}\ \text{J}$$
$$= -13{\cdot}7\ \text{eV}.$$

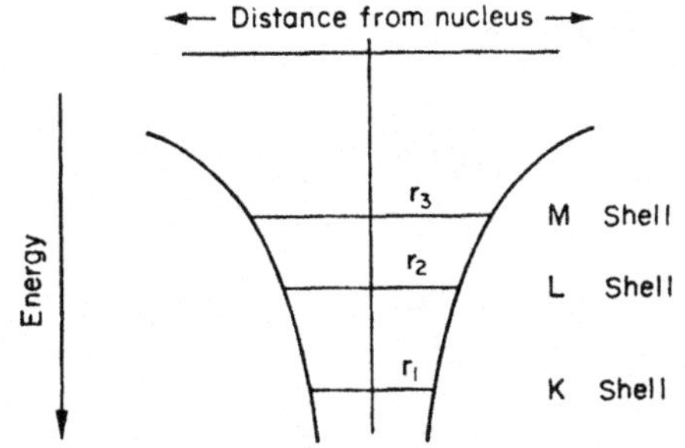

FIG. 1.5. Energy levels for hydrogen atom.

Other values may easily be calculated.

For a more complex atom there are still definite energy levels although these have different values from the above, due to the greater charge on the nucleus. The fact that many electrons have elliptical orbits does not radically affect the energy level diagram.

When a collection of atoms is considered, electrons are influenced by the fields of other atoms as well as the parent atom and the energy levels spread out into bands. There are still permissible energy levels and forbidden energy gaps for a given atomic spacing, as shown in Fig. 1.6. This is a hypothetical diagram but indicates how the single energy levels for isolated atoms form a band for a collection of atoms.

In a solid the complete picture of the energy band diagram is as shown in Fig. 1.7. Discrete energy bands are shown spaced by forbidden energy bands. Spaced through the diagram are dotted curves representing the potential energy $(-e^2/4\pi\epsilon_0 r)$ and at the surface is the energy barrier restraining the electrons within the material.

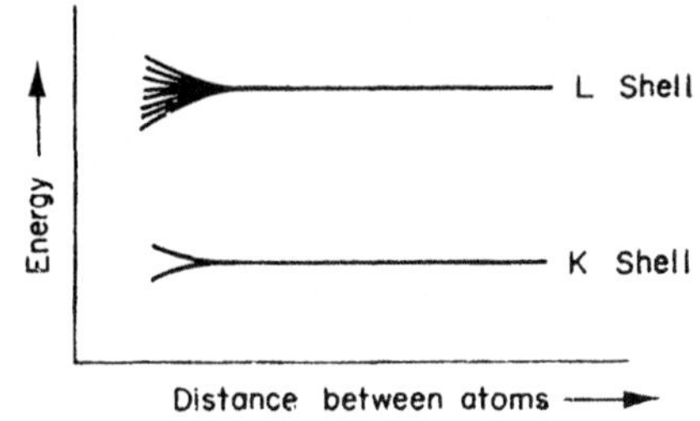

FIG. 1.6. Energy levels and atomic spacing.

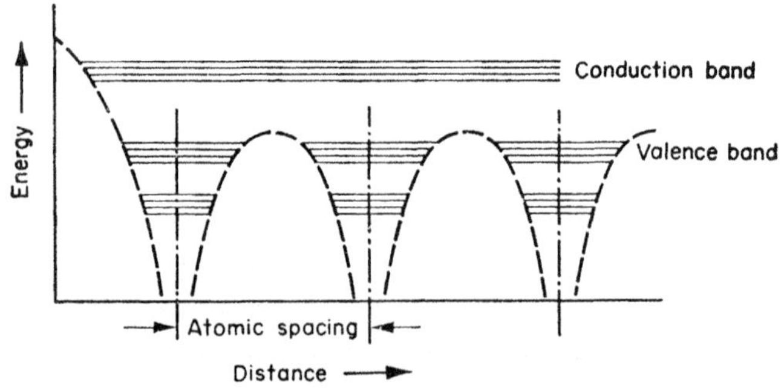

FIG. 1.7. Energy levels in a solid.

The outermost band in which the electrons have sufficient energy to overcome the attractive force of the nucleus contains the electrons available for conduction. These are the free electrons in a conductor and the band is known as the conduction band. The inner bands contain electrons more tightly bound to the atoms and the outermost of these is known as the valency band and it contains valency electrons.

1.6. Fermi Level

If we consider a complete crystal and plot the number of electrons, with a given energy, against energy the envelope is a

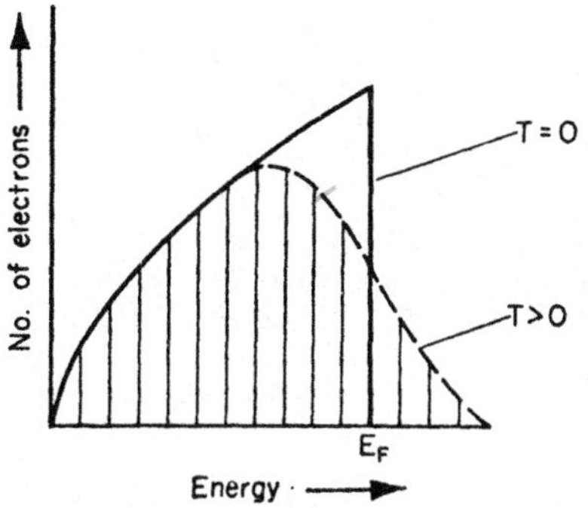

FIG. 1.8. Fermi level.

curve following the Fermi–Dirac distribution law. At absolute zero all energy levels are filled up to an energy level E_F. At any finite temperature some electrons will acquire thermal energy and jump to a higher energy level so that there are as many full energy levels above as there are empty levels below. The Fermi–Dirac distribution law may be stated as

$$f(e) = \frac{1}{1+e^{(E-E_F)/kT}},$$

where f(e) is the probability of finding an electron with energy E, E_F is a particular energy level known as the Fermi level, k is Boltzmann's constant $= 1 \cdot 38 \times 10^{-23}$ J/°K, and T is the absolute temperature (°K).

The Fermi level may be defined as the energy level at which the probability of finding an electron is a half at any finite temperature. It is a measure of the potential energy of the system and is shown graphically in Fig. 1.8.

1.7. Energy Levels in Conductors

The simple explanation of conduction is that free electrons wander through a conductor as shown in Fig. 1.9 to be attracted

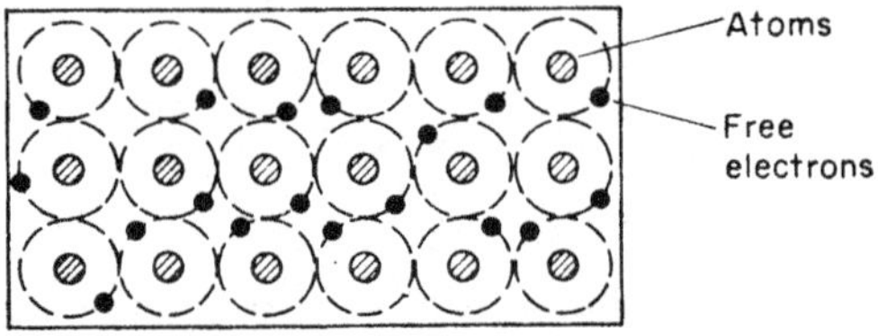

FIG. 1.9. Free electrons in a conductor.

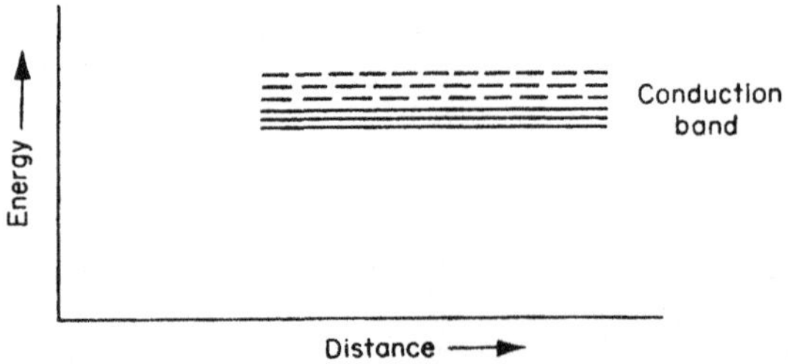

FIG. 1.10. Energy levels in a conductor.

to the positive electrode when a potential is applied. In terms of energy levels the free electrons form a partially filled conduction band as shown in Fig. 1.10. The application of an electric field tilts the conduction band, enabling the electrons to absorb small amounts of energy and rise to a higher energy level, but dropping to a lower level on collision and immediately acquiring more energy. This is shown in Fig. 1.11, the points X representing the points where collisions take place. The energy lost in collisions is used to heat the lattice.

Current flow is due to the irregular motion of electrons through the conductor, that gives a net charge displacement in a definite direction. This type of motion is known as drift and the ratio

$$\frac{\text{Drift velocity } (v_e)}{\text{Applied electric field } (E)}$$

is known as the mobility of the electron (μ_e),

i.e. $$v_e = \mu_e E\,.$$

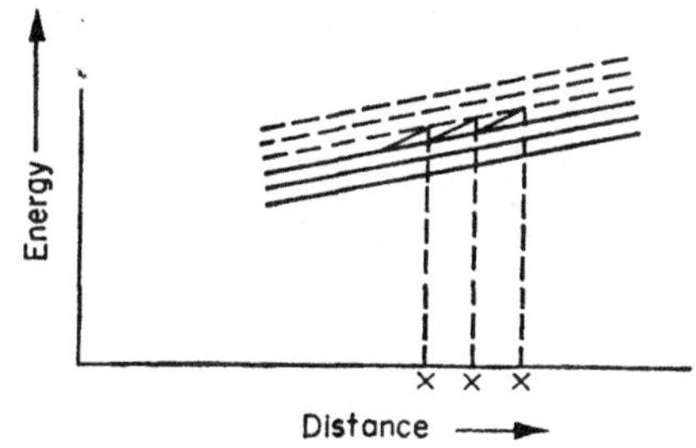

FIG. 1.11. Energy levels in a conductor with an applied field.

If v_e is in m/s, E in V/m the units of μ_e are m^2/V sec.

The value of the current flowing is the charge passing a given point per second,

i.e. $$I = N_e A v_e e\,,$$

where N_e is the number of free electrons per unit volume, A is the area of the conductor, e is the charge on an electron,

i.e. $$I = N_e e \mu_e A E = N_e \mu_e \frac{AV}{L}\,,$$

where V is the potential difference across the conductor, L is the length of the conductor in metres.

Then Resistance $$\frac{V}{I} = \left(\frac{1}{N_e e \mu_e}\right) \cdot \frac{L}{A}\,.$$

$1/N_e e \mu_e$ obviously corresponds to the resistivity ρ.

The increase in resistance with temperature is due to the lower mobility resulting from the increased number of collisions due to the larger atomic vibrations.

1.8. Hall Effect in Conductors

If a conductor carries a current I in a field of flux density B tesla the electrons will experience a deflecting force Bev_e newtons, as shown in Fig. 1.12.

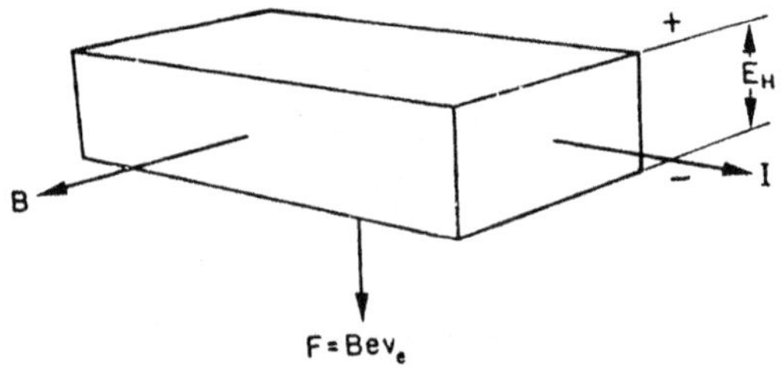

FIG. 1.12. Hall effect.

The accumulation of electrons at the base will give rise to an electric field E_H across the conductor and the production of this field is known as the Hall Effect.

In equilibrium $eE_H = Bev_e$,

i.e.
$$v_e = \frac{E_H}{B},$$

and Current density
$$J = \frac{I}{A} = N_e e v_e$$
$$= \frac{N_e e E_H}{B}.$$

Hence $1/N_e e = E_H/BJ$ and this is known as the Hall Coefficient (R_H). This equation enables N_e to be determined.

Also $\mu_c = 1/N_e e\rho = R_H/\rho$ enabling the mobility to be found.

1.9. Contact Potential between Metals

When two metals A and B having different Fermi levels, as shown in Fig. 1.13, are brought into contact electrons move from the higher energy levels in B to the lower energy unoccupied levels in A. The negative charge acquired by A will raise its energy level diagram while that of B will be lowered. Electrons will flow until the Fermi levels are equal and an energy barrier equal to $E_{F_B} - E_{F_A}$ electron volts will exist between the metals. This corresponds to a potential barrier $E_{F_B} - E_{F_A}/e$ volts, where e is the charge on an electron.

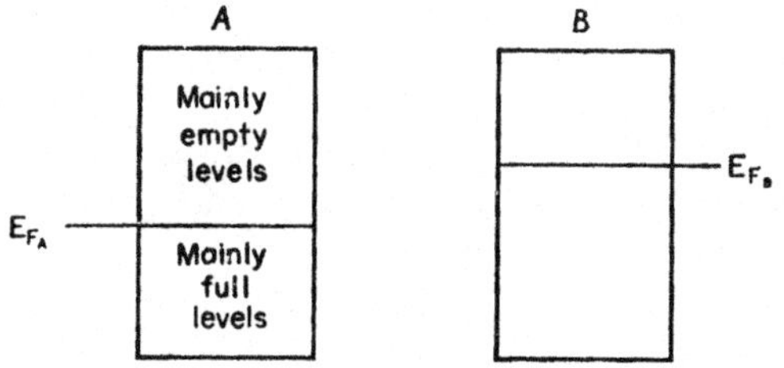

FIG. 1.13. Fermi levels for different conductors.

1.10. Energy Levels in Insulators

The simple picture of insulators is electrons bound tightly to their parent atoms and requiring a large electric field to remove electrons from the attraction of the nuclei.

In terms of energy levels this means a full valence band, an empty conduction band and a large energy gap between them as shown in Fig. 1.14. For conduction to take place electrons must be given sufficient energy to jump from the valence to the conduction band. An increase of temperature enables some electrons to acquire energy to move into the conduction band and this accounts for the negative resistance temperature coefficient of insulators. The production of extra carriers far outweighs the effects of lattice scattering that predominate in the case of conductors.

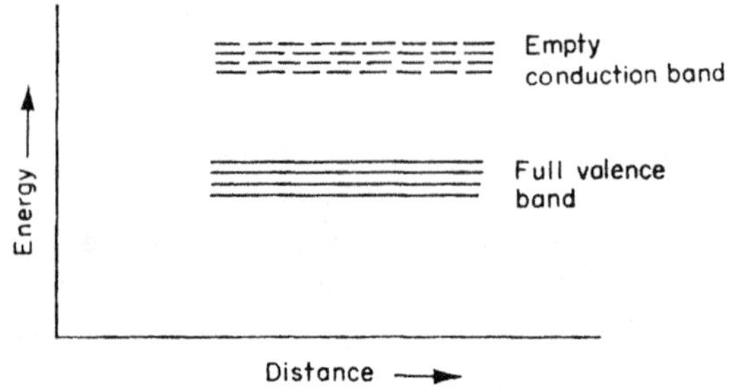

FIG. 1.14. Energy levels in an insulator.

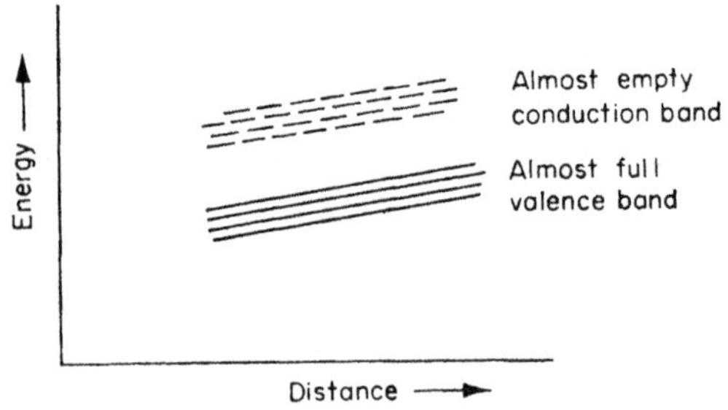

FIG. 1.15. Energy levels in an intrinsic semiconductor.

1.11. Energy Levels in Intrinsic Semiconductors

The simple picture of conduction in an intrinsic semiconductor is the breaking of the covalent bands shown in Fig. 1.2. These will break easier at an elevated temperature and the resistance will decrease with temperature. The energy level diagram of a pure semiconductor is similar to that of an insulator but with a much narrower energy gap, 0·72 eV for germanium and 1·1 eV for silicon. Carbon which has the same structure has an energy gap of 7 eV which puts it in the class of an insulator while tin with an energy gap of 0·1 eV is only suitable at very low temperatures. Certain compounds such as gallium–arsenide and indium–phosphide have a similar energy level diagram but up to the early

1960's have not been used in commercial semiconductor devices to any large extent. At room temperature the energy gap is such that some electrons will acquire sufficient energy to cross from the valence to the conduction band, and with an applied field an electron current will flow. The energy level diagram is shown in Fig. 1.15 and $I_e = n_I e v_e A$. This equation is identical with that for a conductor but v_e refers specifically to the drift velocity of electrons and n_I refers to the number of free electrons in an intrinsic semiconductor.

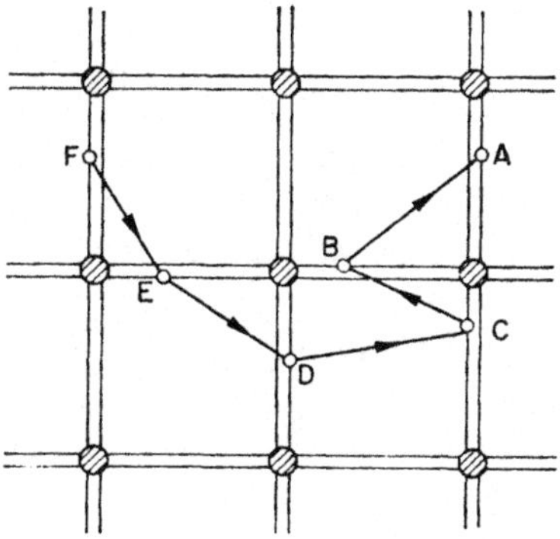

FIG. 1.16. Hole flow in an intrinsic semiconductor.

The movement of electrons from the valence band corresponds to the breaking of covalent bonds and besides the production of a conduction electron, the negative charge carrier, a positive charge carrier will be formed. Assume a bond has been broken at A in Fig. 1.16, the electron will have moved through the lattice leaving a hole in the covalent bond. An electron at B may jump into the bond at A, and an electron at C may jump into the hole at B. In this way by a succession of electron movements a hole will appear at F and a negative charge will have moved from F to A. It is more convenient to regard a positive charge to have moved from A to F and this conception gives rise to a hole as a positive charge carrier. These holes are due to the movement of electrons in the valence band and each electron movement corresponds to a collision so that the drift velocity of holes is less than that of electrons.

The current due to the hole movement is given by

$$I_h = p_I e v_h A,$$

where v_h is the drift velocity of the holes and p_I is the number of holes in an intrinsic semiconductor.

The total current flow is then the sum of that due to electrons and holes.

i.e.

$$I = n_I e(v_e + v_h)A$$
$$= n_I e(\mu_e + \mu_h)AE,$$

where $n_I = p_I$, since for an intrinsic semiconductor the number of holes and electrons is the same, and μ_e and μ_h are the mobilities of electrons and holes respectively.

Typical values for germanium are $n_I = 2 \times 10^{19}$ charge carriers/m^3, $\mu_e = 0{\cdot}36$ m^2/V sec and $\mu_h = 0{\cdot}17$ m^2/V sec. The corresponding values for silicon are $n_I = 1{\cdot}5 \times 10^{16}$/m^3, $\mu_e = 0{\cdot}12$ m^2/V sec and $\mu_h = 0{\cdot}025$ m^2/V sec.

Using these values for germanium:

$$I = 2 \times 10^{19} \times 1{\cdot}6 \times 10^{-19} \quad (0{\cdot}53)AV/L.$$

For unit dimensions:

$$\text{Resistivity} = \frac{V}{I} = \frac{1}{3{\cdot}2 \times 0{\cdot}53} = 0{\cdot}6\ \Omega\text{-m}.$$

While for silicon:

$$\text{Resistivity} = 3000\ \Omega\text{-m}.$$

The above values are for 300°K and n_I increases with temperature while μ_e and μ_h decrease but the net effect is a decrease in resistivity.

1.12. Energy Levels in Extrinsic Semiconductors

The introduction of pentavalent or trivalent impurities in intrinsic germanium or silicon in very small amounts, normally about 1 part in 10^8, has a large effect on the conductivity.

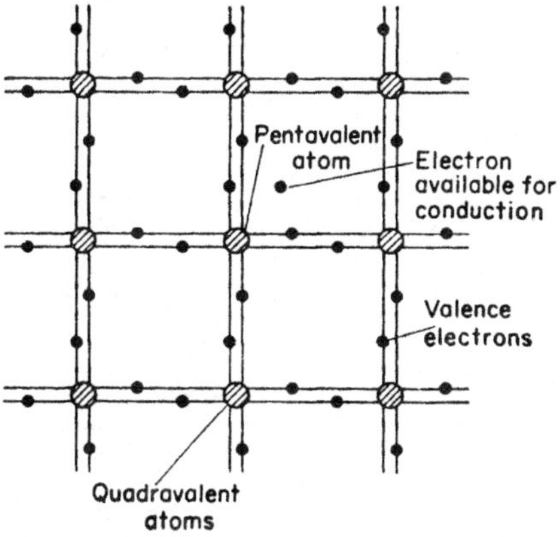

FIG. 1.17. Covalent bonds in an *n* type semiconductor.

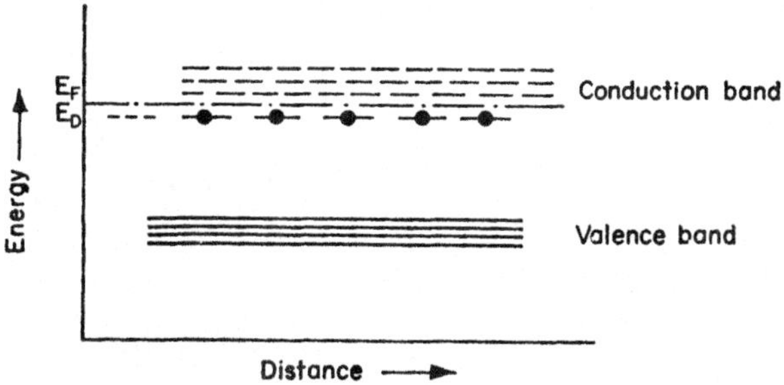

FIG. 1.18. Energy levels in an *n* type semiconductor.

The added impurities or doping elements give rise to electrons outside or holes within the covalent bonds, and the case of a pentavalent impurity is shown simply in Fig. 1.17. Four of the five valence electrons are absorbed in the covalent bonds between the atoms but the fifth electron, although loosely bound to its parent atom, is easily detached by the application of an electric field, or an increase in thermal energy.

In terms of energy levels the loosely bound fifth electron has an energy level just below the conduction band as shown in Fig. 1.18.

The pentavalent impurities such as phosphorus, arsenic or antimony that provide an electron for conduction are known as donors and the local energy level as the donor level. No holes are produced and conduction takes place by raising electrons from the

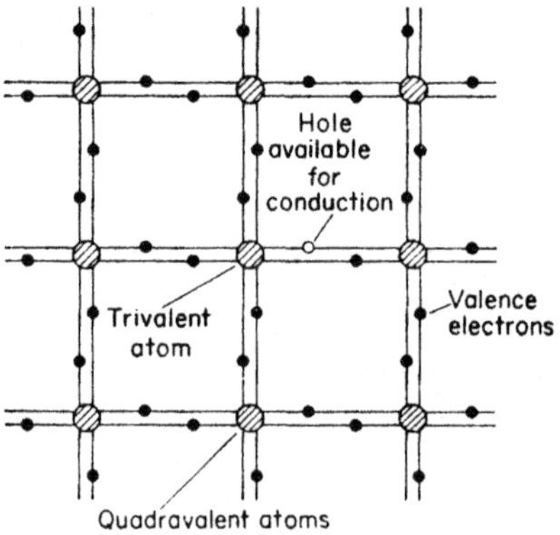

FIG. 1.19. Covalent bonds in a p type semiconductor.

local donor level into the conduction band. The material is known as n type since the current flow is due to negatively charged electrons. In practice the donor level is 0·01 eV below the conduction band for germanium and 0·054 eV for silicon and at normal temperatures the Fermi level may be regarded as lying between the conduction band and the donor level. The Fermi level does vary considerably with temperature and the reader is recommended to one of the many physics books available for a full treatment of this subject.

In the case of a trivalent impurity such as boron, aluminium, gallium or indium, the three valence electrons form bonds, but one bond is left incomplete and gives rise to a hole. As in the case of an intrinsic semiconductor the application of an applied field causes the hole to move through the crystal lattice and conduction is by means of the positively charged holes. Free electrons are not produced in this process and the material is known as p type. This is shown in Fig. 1.19 while Fig. 1.20 shows the corresponding energy level diagram. The hole gives rise to a local empty energy level just above the valence band. Conduction is by means of hole movement at the top of the valence band, the local empty energy level readily accepting electrons from the valence band. The impurity energy level introduced is known as the acceptor energy

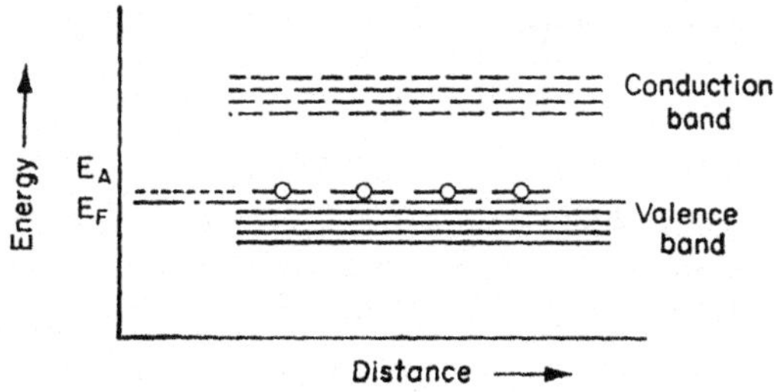

FIG. 1.20. Energy levels in a p type semiconductor.

level E_A and the Fermi level at normal temperature lies between the top of the valence band and the acceptor energy level. As in the case of donor impurities the reader is recommended to further reading on this subject.

The Hall effect is more noticeable with extrinsic semiconductors than with conductors, but in this case the Hall coefficient R_H is usually taken as $\frac{3}{8}\pi \cdot (1/n_n e)$ m^3/C for an n type semiconductor and $-\frac{3}{8}\pi \cdot (1/p_p e)$ m^3/C for a p type semiconductor. In these expressions n_n denotes the number of " free " electrons in an n type semiconductor and p_p the number of " free " holes in a p type semiconductor. The factor $\frac{3}{8}\pi$ is due to a different distribution of velocities of the carriers in a semiconductor, but must be modified for large values of magnetic field strength.

The calculation of resistivity of extrinsic semiconductors involves a knowledge of the number of charge carriers after doping and to calculate this the Law of Mass Action must be used.

For a donor impurity this may be stated symbolically as

$$n_n p_n = n_I p_I = n_I^2, \quad \text{since} \quad n_I = p_I,$$

where n_n and p_n are the number of " free " electrons and holes in the n type semiconductor after doping and n_I and p_I the number of electron and hole carriers before doping. Stated in another way, at constant temperature, the product of the number of electron carriers and the number of hole carriers is independent of the density of donor centres. Physically, the introduction of an n type impurity fills some of the holes produced by thermal action.

For an acceptor impurity

$$n_p p_p = n_I p_I = n_I^2,$$

where n_p and p_p are the number of " free " electrons and holes in the p type semiconductor. The introduction of the p type impurity absorbs some of the electrons produced by thermal action.

If we take the case of germanium stated previously

$$(n_I = 2\times 10^{19} \text{ charge carriers/m}^3),$$

the addition of 10^{20} atoms/m^3 of indium, a trivalent impurity, gives

$$n_p p_p = 4\times 10^{38} \quad \text{and} \quad p_p - n_p = 10^{20}.$$

Therefore $$p_p - 4\times 10^{38}/p_p = 10^{20}$$

or $$p_p^2 - 10^{20} p_p - 4\times 10^{38} = 0.$$

Solving and taking the positive root gives

$$p_p = 1{\cdot}04\times 10^{20} \quad \text{and} \quad n_p = 0{\cdot}04\times 10^{20}.$$

Therefore $$I = e(n_p\mu_e + p_p\mu_h)AE \text{ amps}$$

and resistivity

$$\rho = \frac{1}{e(n_p\mu_e + p_p\mu_h)}\ \Omega\text{-m}$$

$$= \frac{1}{1{\cdot}6\times 10^{-19}(0{\cdot}04\times 10^{20}\times 0{\cdot}36 + 1{\cdot}04\times 0{\cdot}17\times 10^{20})}$$

$$= 0{\cdot}327\ \Omega\text{-m}.$$

The addition of 2×10^{20} atoms/m^3 of antimony, a pentavalent impurity that has a greater effect than the original trivalent impurity, gives

$$n_n - p_n = N_D - N_A = 10^{20},$$

where N_D and N_A is the number of donor and acceptor atoms added per cubic metre.

Then since $n_n p_n = 4 \times 10^{38}$, solving as before gives $n_n = 1{\cdot}04 \times 10^{20}$ and $p_n = 0{\cdot}04 \times 10^{20}$ and a resistivity of 0·164 Ω-m.

Questions for Chapter 1

Q.1.1. Calculate the radii and potential energy levels of the first two possible orbits for an ionized helium atom.

Q.1.2. The number of free electrons in copper is $1{\cdot}1 \times 10^{29}$/m^3 and its resistivity is 10^{-8} Ω-m. Determine the mobility of the electrons and the Hall coefficient for copper. Hence determine the current density which gives a Hall voltage of 1 μV in a field of 1 Wb/m^2.

Q.1.3. Assuming that the resistivity of pure silicon is 3000 Ω-m and the mobilities of electrons and holes are 0·12 m^2/V sec and 0·025 m^2/V sec respectively, determine the resistivity of a sample of silicon when 10^{19} atoms of phosphorus are added per cubic metre. If 2×10^{19} atoms of boron are now added what is the resultant resistivity? What would the answers be if germanium were the base material? Assume the resistivity of pure germanium is 0·6 Ω-m, electron mobility is 0·36 m^2/V sec and hole mobility is 0·17 m^2/V sec.

Q.1.4. A square of *n* type germanium whose side is 1 cm is 0·025 cm thick. A current of 2·5 mA from a 1·35 V supply is passed through the slice which is placed in a field of flux density 0·05 tesla. If the Hall voltage developed across the specimen is 20 mV, determine the number of free electrons per cubic metre, the Hall coefficient, the resistivity and the electron mobility.

Q.1.5. What is meant by energy level of an electron in an atom? With reference to solids explain the following terms, illustrating your explanation with diagrams where this is possible: filled energy band, conduction band, Fermi level, crystal lattice, valency electrons, covalent bond, intrinsic and impurity semiconductors. (I.E.E., June 1960)

CHAPTER 2

Physics of Semiconductor Devices

THE physical principles embodied in the first chapter are applied in this to practical semiconductor devices. Diodes and transistors are considered in detail but the extension of the theory to four layer and other special devices is left to the chapters particularly associated with such units.

2.1. The *pn* Junction

In multi-layer devices there are two components of the current to be considered. One is the drift current, mentioned in the previous chapter, and the other the diffusion current due to the movement of carriers from regions of high concentration to regions of lower concentration, without the application of an applied field.

The rectifying action of a *pn* junction depends on the flow of a diffusion current. Free electrons, from the *n* type region, and free holes, from the *p* type region, diffuse across the junction, giving rise to a depletion layer in the neighbourhood of the junction. The depletion layer is a region where holes and electrons recombine, so that there are no free charge carriers, and it acts as a dielectric. Diffusion of electrons into the *p* region and holes into the *n* means there is charge transfer which continues until the charge built up is sufficient to stop further movement of carriers. The voltage developed across the depletion layer (V_{dep}) is shown in Fig. 2.1 and its value is given by q/C_{dep}, where C_{dep} is the capacitance of the depletion layer. Minority carriers (i.e. electrons in the p type and holes in the n type) are continually being produced thermally and swept across the junction. These are immediately balanced by the flow of majority carriers maintaining the barrier at a constant level.

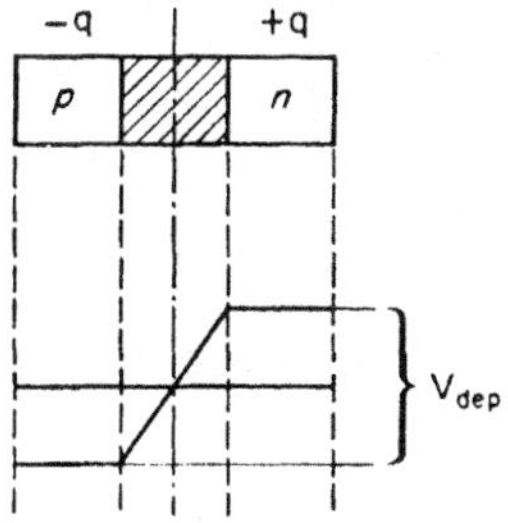

FIG. 2.1. Charge distribution in a *pn* diode.

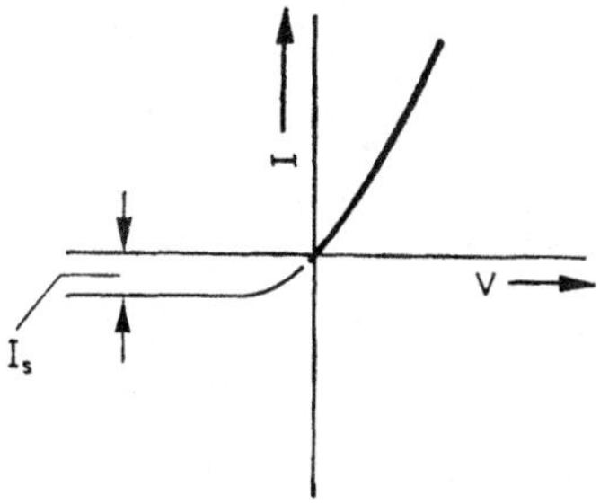

FIG. 2.2. Characteristic of a *pn* diode.

An applied potential will either increase or decrease the potential barrier. If the p type is made positive with respect to the n type, majority carriers will flow to try to maintain the barrier potential. This forward current rapidly increases according to the law:

$$I = I_S(\epsilon^{eV/KT} - 1),$$

where e is the charge on an electron, V is the forward voltage, K is Boltzmann's constant, T is the absolute temperature and I_S is the reverse saturation current.

The reverse current which flows even though the potential barrier is increased is due to minority carriers produced thermally and rapidly approaches its saturation value I_S as shown in Fig. 2.2. A better explanation of the operation of a semiconductor diode is obtained by reference to the energy level diagrams shown in the previous chapter.

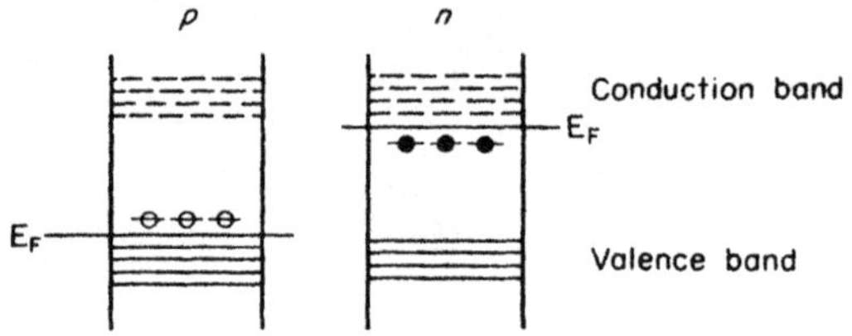

FIG. 2.3. Energy levels in *p* and *n* type semiconductors.

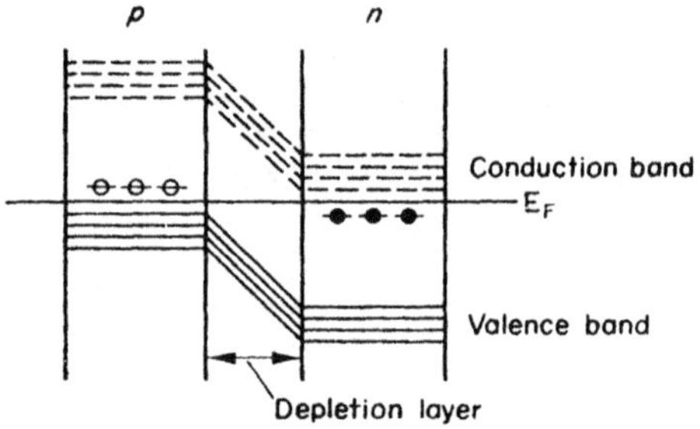

FIG. 2.4. Energy levels in a *pn* diode with zero bias.

2.2. *pn* Junction with Zero Bias

The energy level diagrams shown in the previous chapter are reproduced in Fig. 2.3. When a *pn* junction is formed electrons flow from donor levels in the *n* type to the unoccupied acceptor levels in the *p* type giving rise to a region of intrinsic semiconductor with no free charge carriers. This is the depletion layer. Electrons flow until the raising of the energy level diagram of the negatively charged *p* type coupled with a fall in the energy level diagram of the positively charged *n* type produces an energy barrier sufficient to stop further diffusion of electrons. This equilibrium condition corresponds to the Fermi levels being equal as shown in Fig. 2.4.

2.3. *pn* Junction with Forward Bias

With the *p* side of the junction made positive with respect to *n*, the energy barrier is lowered by an amount eV_F joules, where e is

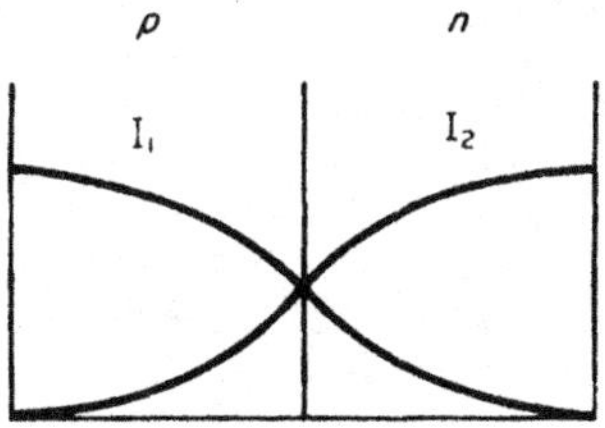

FIG. 2.5. Hole and electron flow in a *pn* diode.

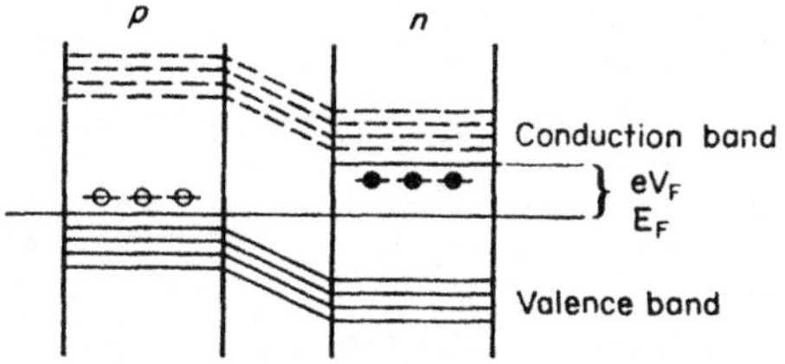

FIG. 2.6. Energy levels in a *pn* diode with forward bias.

the charge on an electron and V_F the voltage applied. This results in the flow of majority carriers, electrons from the *n* and holes from the *p* type semiconductors, giving rise to an appreciable forward current. The current carriers recombine after crossing the junction and the distribution of current carriers is indicated in Fig. 2.5, where I_1 and I_2 are the currents carried by holes and electrons respectively.

The movement of electrons from the negative electrode and holes from the positive electrode results in a narrowing of the depletion layer and the energy level diagram for a forward biased *pn* junction is shown in Fig. 2.6.

2.4. *pn* Junction with Reverse Bias

If the *p* side is made negative with respect to the *n* the energy barrier is increased by an amount eV_R joules, where V_R is the reverse voltage applied. The only current that flows is that due to

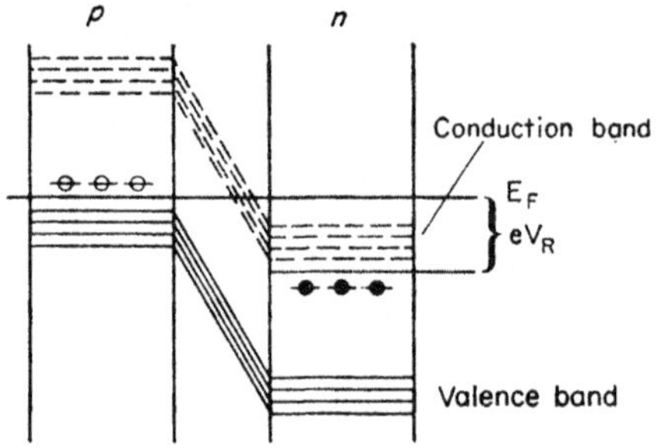

FIG. 2.7. Energy levels in a *pn* diode with reverse bias.

minority carriers produced thermally and is known as the leakage current I_S. Electrons are attracted to the positive potential and holes to the negative potential resulting in a widening of the depletion layer as shown in Fig. 2.7.

The above considerations lead to a *pn* diode characteristic as shown in Fig. 2.2. The equation $I = I_S(e^{eV/kT} - 1)$, however, is not valid for high forward currents when large charge concentrations are involved or high reverse voltages when breakdown occurs.

2.5. The Zener Diode

The term Zener Diode is used commercially to cover the whole range of stabilizing diodes, whose operation depends on the reverse breakdown characteristic. This is in fact a misnomer since only those with a breakdown voltage of less than about 6 V rely on the Zener Effect for their operation, while higher voltage diodes depend on the avalanche effect.

When a *pn* diode is reverse biased all the applied voltage appears across the depletion region, since it has no free charge carriers and hence an infinite resistance. If the impurity content is about 1 part in 10^5, the relatively small dimensions of the depletion layer result in a very high field strength in this region. This causes a rapid increase of current at low voltages due to the breaking of covalent bonds and is known as the zener effect. Zener diodes using this mode of operation are usually made of silicon with a zener breakdown voltage lying between -3 V and -6 V, and a negative temperature coefficient. Higher voltage stabilizing diodes utilize the avalanche effect, and have a much lower impurity content. The few free electrons acquire sufficient kinetic energy from the applied field to break the covalent bonds, and the electrons so produced are themselves accelerated. This cumulative process leads to a rapid increase in current and may be compared to ionization in a gas-filled valve. Such diodes have a positive temperature coefficient. It is difficult to separate the two effects in a practical Zener diode.

2.6. The Tunnel (Esaki) Diode

This is a *pn* diode with a very high degree of doping (1 part in 10^3 or 10^4) such that an abrupt *pn* junction is formed with a very narrow depletion layer, about 10^{-6} cm (10^{-2} micron) wide. The characteristic of such a device is shown in Fig. 2.8 and it may be explained by reference to the energy level diagrams shown in Figs. 2.9, 2.10 and 2.11.

The high degree of doping does not produce the isolated acceptor and donor levels as in the conventional *pn* diode but a band of such levels. These bands of donor and acceptor levels overlap the conduction and valence bands, effectively increasing the number of energy levels in these bands. The Fermi level in the *p* type now lies inside the valence band and in the *n* type inside the conduction band. Equilibrium conditions for zero bias are shown in Fig. 2.9. The Fermi levels are equal with empty acceptor levels on the p side and full donor levels on the n side.

A small forward bias V_A, corresponding to the point A on the characteristic, gives rise to the energy level diagram shown in

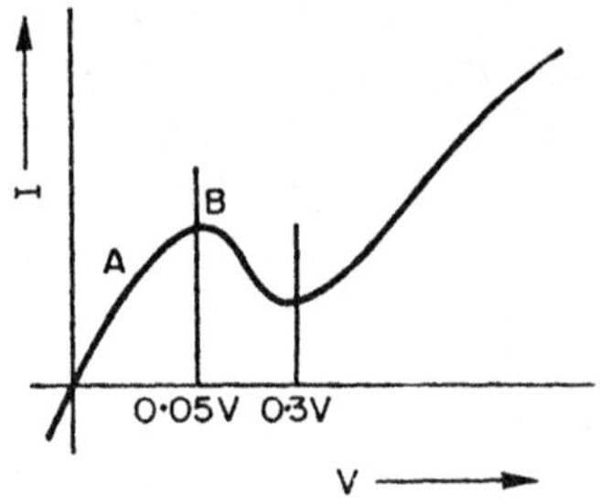

FIG. 2.8. Characteristic of a tunnel diode.

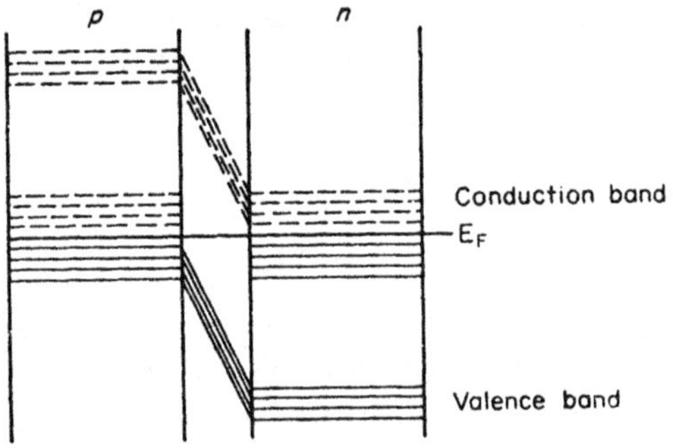

FIG. 2.9. Energy levels in a tunnel diode with zero bias.

Fig. 2.10. The energy levels on the n side are raised with respect to those on the p side and full donor levels are opposite empty acceptor levels with a very narrow depletion layer between them. Electrons can "tunnel" through the very narrow depletion region and give rise to a current of a few milliamperes where the normal diffusion current would be negligible. This movement of electrons is possible when the energy barrier is too high for the electrons to surmount and takes place at approximately the speed of light, enabling the device to be used at microwave frequencies.

A larger forward bias V_B corresponding to the point B on the characteristic gives rise to the energy level diagram shown in Fig.

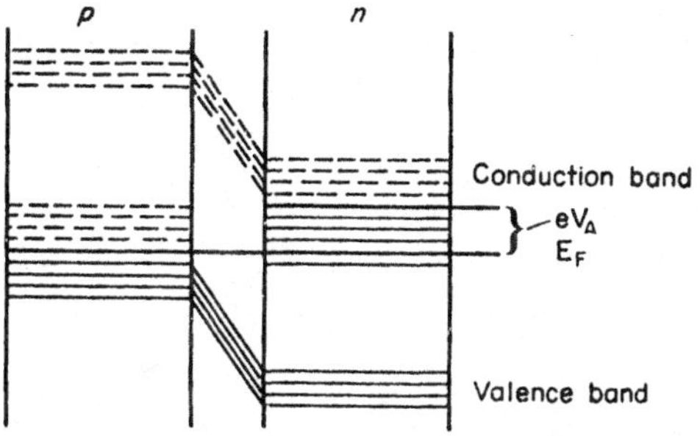

FIG. 2.10. Energy levels in a tunnel diode with small forward bias.

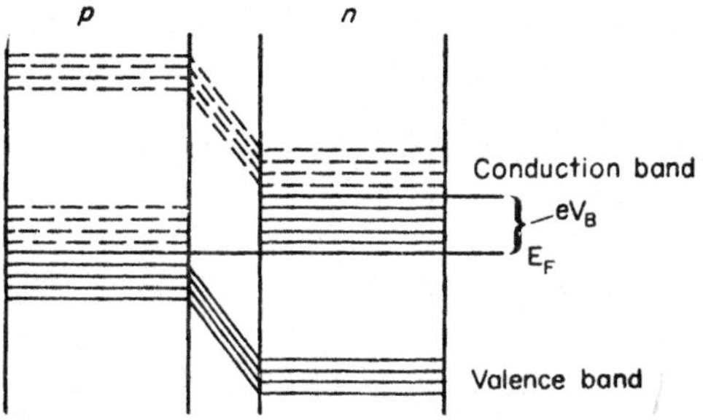

FIG. 2.11. Energy levels in a tunnel diode with critical forward bias.

2.11. This is the condition where the maximum number of full donor levels are separated from the maximum number of empty acceptor levels by the narrow depletion layer and corresponds to maximum " tunnel " current.

Further increase in forward voltages reduces the " tunnel " current and at a few hundred millivolts it is negligible compared with the diffusion current. The resultant current flowing through a tunnel diode is then a combination of " tunnel " and diffusion currents giving rise to the characteristic as shown in Fig. 2.12.

Due to the high impurity content, reverse (zener) breakdown occurs at a very low value and normal amplifying and oscillating circuits do not utilize this part of the characteristic.

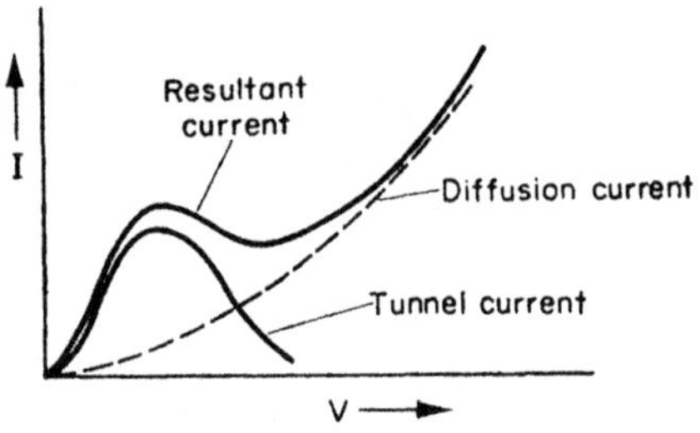

FIG. 2.12. Diffusion and " tunnel " currents in a tunnel diode.

2.7. Metal–Semiconductor Diodes

These are the oldest types of semiconductor devices, and were used for years before the theory of operation was fully understood. Copper-oxide and selenium rectifiers are still used extensively, but are being replaced in many applications by *pn* diodes.

The simple theory assumes a barrier layer between metal and semiconductor, e.g. copper and copper-oxide, with easy electron flow across the barrier layer from metal to semiconductor, but difficult electron flow in the opposite direction.

The interpretation of the operation in terms of energy levels is difficult since it relies on the action of impurities at the surface of the semiconductor. Covalent bonds are incomplete at the surface of a semiconductor and combine with impurities. This gives rise to a range of energy levels within the forbidden gap and the condition for an *n* type semiconductor is shown in Fig. 2.13. Electrons move from the donor levels and valence band into these surface levels until sufficient surface charge builds up to stop further diffusion from the interior of the semiconductor. The holes produced in the valence band give rise to a *p* type region slightly below the surface as shown in Fig. 2.14.

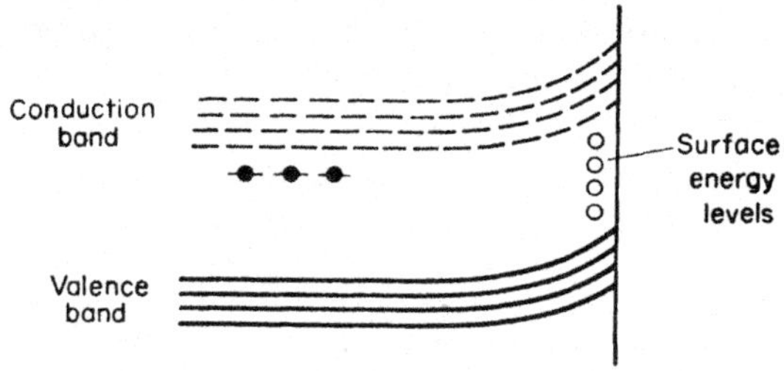

FIG. 2.13. Energy levels at the surface of an *n* type semiconductor.

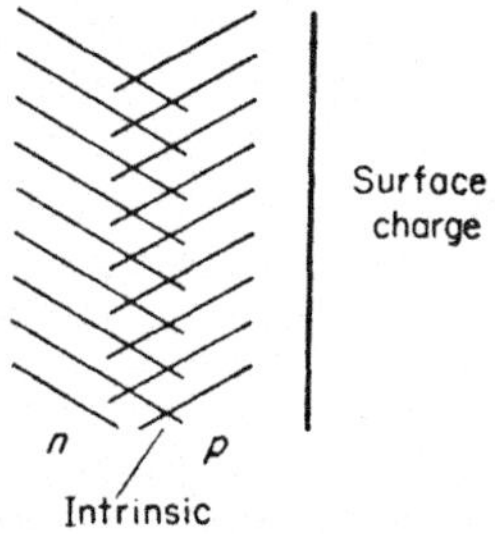

FIG. 2.14. Rectifying junction formed by surface impurities.

A corresponding energy level diagram is produced at a metal–semiconductor junction and the *pn* junction so formed rectifies in a similar manner to that described previously. An *n* type semiconductor acts in the manner described, but copper oxide and selenium are *p* type semiconductors and in this case the impurities produce an *n* type region near the surface.

2.8. The *pnp* Transistor

The *pnp* transistor may be regarded as two *pn* diodes back to back, the first, which is forward biased, providing the minority carriers for the second reverse biased diode.

The construction of the transistor, shown in Fig. 2.15 and described in detail in the next chapter, is such that the impurity content in the narrow *n* type region is very much less than in the *p* type regions. This means that in the forward biased input circuit, the flow of current carriers across the junction is mainly in the form of holes from the *p* to *n* regions. Some of these holes, which

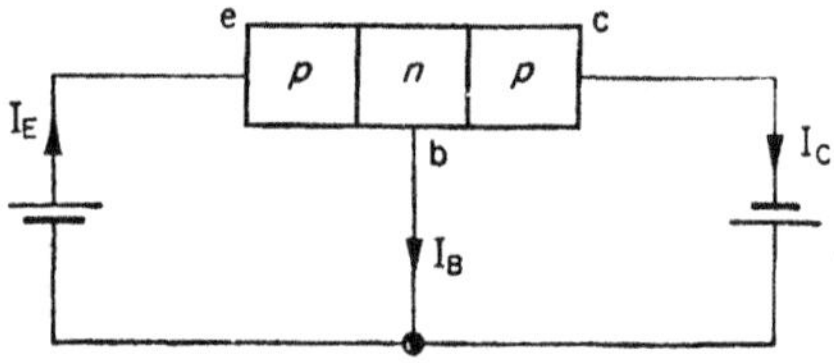

FIG. 2.15. The *pnp* transistor.

are minority carriers as far as the *n* type region is concerned, recombine with electrons but most diffuse across the base and flow into the high impedance output circuit.

A small electron flow results from the few free electrons in the base so that the total input current is due to holes and electrons. Only the holes contribute to the output current and this is usually stated in terms of an emitter efficiency which is the ratio of hole input current to total input current.

The input direct current I_E flows as electrons in the connection to the positive pole of the battery and as these are drawn out of the *p* type region, holes are injected and flow towards the junction. This region is known as the emitter and the second *p* region the collector, while the narrow *n* type region of relatively high resistivity is designated the base. The output current I_C is less than I_E due to recombination in the base and this loss of negative charge leads to a small electron current I_B flowing into the base region.

One factor that has not so far been considered is the leakage current that flows in the reverse biased collector diode. The relation between I_C and I_E is shown in Fig. 2.16. I_{CO} represents the leakage current that flows even when I_E is zero and the diagram shows that I_C increases linearly with I_E.

These considerations lead to the two basic expressions that are used in the design and solution of circuits as far as direct currents are concerned, i.e.

$$I_E = I_B + I_C,$$
$$I_C = \bar{\alpha} I_E + I_{CO}.$$

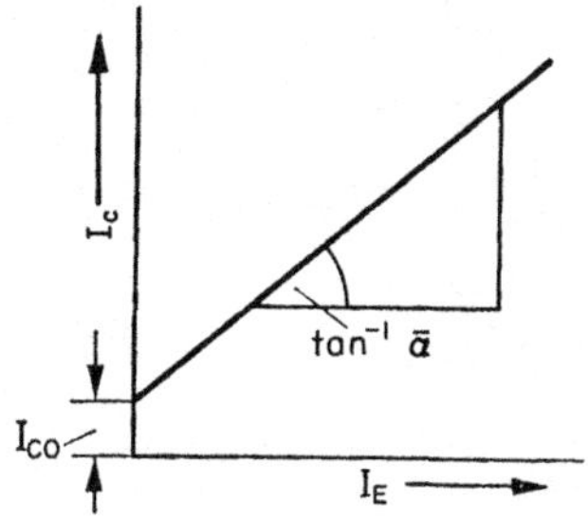

FIG. 2.16. Current relationship for a *pnp* transistor.

Small changes in input current require little power but, due to the high impedance output circuit, considerable variation in output power may be obtained. This leads to the use of transistors as amplifiers, oscillators and other circuit devices as described in later chapters.

The arrangement shown in Fig. 2.15 is known as the common or grounded base configuration, since the base is common to both input and output circuits. The operation may be explained in terms of the energy level diagrams shown in Figs. 2.17 and 2.18. Figure 2.17 shows the transistor with zero bias. Initially, depletion layers are formed in the same manner as diodes and, except for minority carriers produced thermally, no holes or electrons can flow due to the energy barriers set up.

Figure 2.18 shows the transistor with normal bias, the emitter diode being forward biased and the collector diode reverse biased.

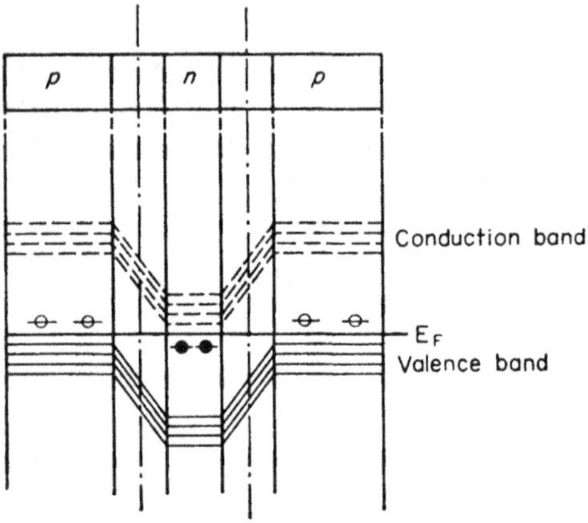

FIG. 2.17. Energy levels in a *pnp* transistor with zero bias.

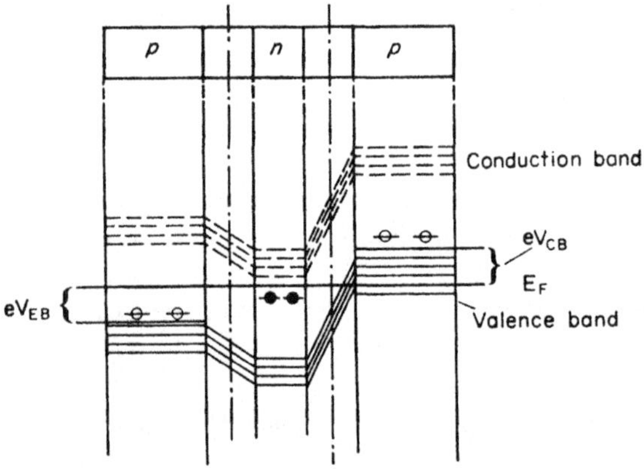

FIG. 2.18. Energy levels in a *pnp* transistor with normal bias.

The energy barrier at the first junction is lowered, allowing the flow of holes which diffuse across the base before being swept across the collector depletion layer.

The other possible circuit configurations, when either the emitter or collector is common to input and output circuits, are known as common emitter or common collector circuits and are shown in Fig. 2.19. In each case the energy level diagram of the n type base is raised with respect to the emitter allowing the flow of emitter and hence collector current.

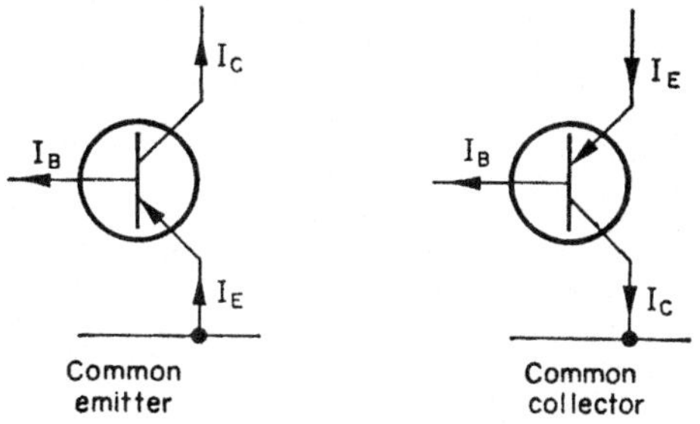

FIG. 2.19. Common emitter and common collector circuits.

The basic d.c. equations for the common emitter may then be written

$$I_B = I_E - I_C,$$

$$I_C = \bar{\alpha} I_E + I_{CO} = \bar{\alpha}(I_B + I_C) + I_{CO},$$

i.e.

$$I_C(1-\bar{\alpha}) = \bar{\alpha} I_B + I_{CO}$$

or

$$I_C = \frac{\bar{\alpha}}{1-\bar{\alpha}} I_B + \frac{I_{CO}}{1-\bar{\alpha}},$$

$$I_C = \bar{\beta} I_B + I'_{CO},$$

where $\bar{\beta} = \bar{\alpha}/(1-\bar{\alpha})$ and $I'_{CO} = I_{CO}/(1-\bar{\alpha}) = I_{CO}(1+\bar{\beta})$.

The last expression indicates the much greater effect of leakage current in this configuration and this is dealt with in detail in later chapters.

Similar d.c. expressions may be developed for the grounded collector configuration giving rise to the equation

$$I_E = \bar{\gamma} I_B + I'_{CO}, \quad \text{where } \bar{\gamma} = 1/(1-\bar{\alpha}).$$

2.9. The *npn* Transistor

This is similar to the *pnp* with a forward biased emitter diode and reverse biased collector diode and is shown in Fig. 2.20. The *p* and *n* regions, however, are interchanged and current flow is by

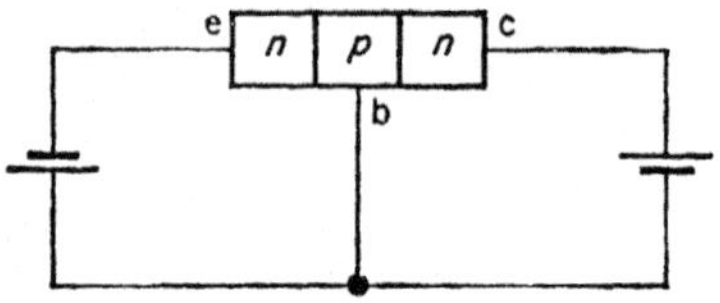

FIG. 2.20. The *npn* transistor.

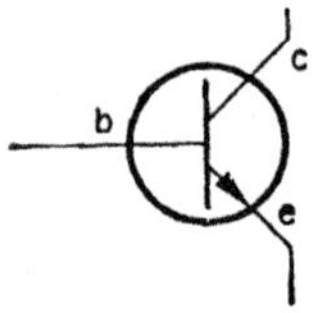

FIG. 2.21.

means of free electrons. The symbol used is shown in Fig. 2.21 and, due to electrons having a higher mobility than holes, such devices can operate at a higher frequency than *pnp* types.

Questions for Chapter 2

Q.2.1. Using as many as possible of the terms listed below write an essay on Semiconductors, Junction Diodes and Transistors.

Intrinsic Semiconductor. Impurity Semiconductor. Donor. Acceptor. *p*-type. *n*-type. Base. Emitter. Collector. Minority Carrier. Lifetime. Depletion Layer. Short Circuit Current Gain. (I.E.E., Nov. 1961)

Q.2.2. Explain the difference between intrinsic, *p*-type and *n*-type conduction in a semiconductor such as germanium.

Describe the mechanism of rectification at a *pn* junction. What factors determine the reverse current in such a rectifier? (B.Sc.(Eng.), 1958)

Q.2.3. Explain using the energy band theory of solids the operation of a *npn* transistor connected in the grounded base configuration.

Q.2.4. Discuss the effects of the amount of impurity added on the characteristics of *pn* diodes. Both the forward and reverse characteristics should be considered.

CHAPTER 3

Construction and Characteristics of Transistors

3.1. Introduction

In this chapter we shall consider some methods of transistor construction and the resulting characteristics. The four general processes which are described in the first part of this chapter are not the only methods available for the production of a *pn* junction, but represent those considered most important at the present time. Devices other than transistors, such as junction diodes and four-layer controlled rectifiers, are fabricated in ways very similar to those available for transistors, and are therefore not considered separately in this chapter. Appropriate details of the characteristics of some of these other devices are given in Chapters 5 and 10.

3.2. Preparation of Crystal

In order to achieve as large a current gain as possible within a transistor the lifetime of the majority charge carrier, in the base region, must be relatively long. This requires that the transistor be made from a single crystal of high quality. At the same time, to obtain other desirable transistor properties, the impurity content of the crystal should be accurately known. Control of the impurity content is best achieved by refining the semiconductor material to maximum possible purity, and then adding impurities under carefully controlled conditions.

Although certain compounds such as indium arsinide have been used, the majority of semiconductor devices are manufactured from germanium or silicon. Mainly by reason of the lower melting point of 950°C, germanium is easier to purify. Silicon is more chemically reactive and has a higher melting point of 1400°C. On the other hand, the wider energy gap in silicon of 1·1 eV (compared with 0·72 eV for germanium) implies a lower leakage current and enables a silicon device to operate with a junction temperature of up to 200°C, whilst germanium junctions must be kept below 100°C.

In the preparation of the semiconductor material, chemical refining is used to reduce the impurity concentration to about 1 part in 10^6. This is then followed by a metallurgical method to

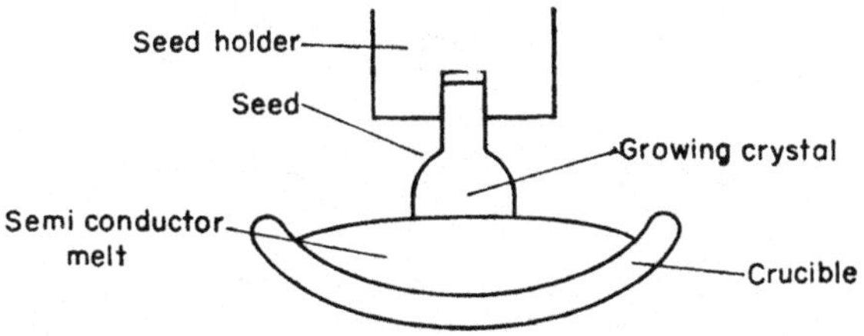

FIG. 3.1. Growth of a single crystal from a melt.

improve this concentration to 1 part in 10^{10}. Silicon may be usefully made more pure than germanium due to the greater energy gap.

The second-stage refining process depends on the principle that, if some material is melted and then frozen from one end, the impurity gradient varies uniformly along the length of the recrystallized material. In the process known as Zone Refining a long ingot of semiconductor material is drawn through a series of radio frequency heating coils. As the molten zones, of about an inch long, traverse the length of this ingot, so the impurities are swept to one end or the other.

When some of the semiconductor is sufficiently pure it may be made into a single crystal with a desired impurity proportion. One method of doing this is to melt the pure semiconductor, in contact with a seed crystal (which is cut from a piece of polycrystalline material) in a crucible, to which the required amount of impurity material is added. As it cools, the semiconductor which is nearest to the seed grows on to that seed to enlarge the single crystal. For good results this must be done under carefully controlled conditions. The crystal is pulled out of the melt at a closely specified speed and melt temperature, with the melt being stirred by rotation of the growing crystal. The basic principles of this process are illustrated in Fig. 3.1.

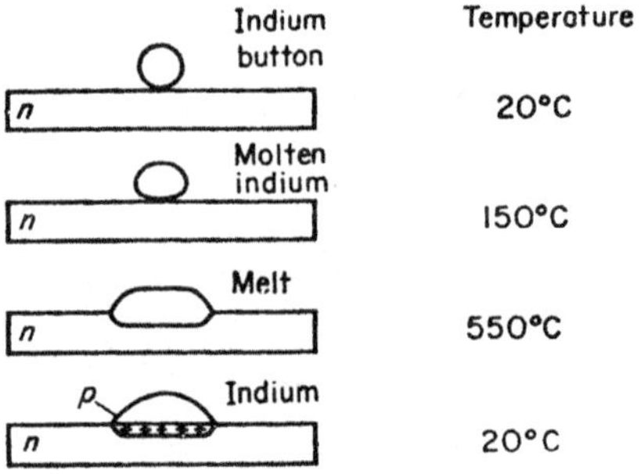

FIG. 3.2. Steps in the production of an alloyed junction.

3.3. Alloying

During the first decade from 1950 most transistors were made by the alloyed junction process. Common examples are of indium alloyed into germanium or aluminium into silicon, to form *pnp* transistors. In the manufacture of germanium alloyed transistors an *n* type crystal is usually cut into small wafers of 2 mm square and 0·1 mm thick. The steps in the production of one junction are illustrated in Fig. 3.2. When the indium melts it dissolves an amount of germanium determined by the volume and temperature of the indium. Then, as the structure cools, the germanium recrystallizes, but now contains some indium in the crystal structure and it is therefore *p* type. As this type of junction formation leads to heavy doping, the base region needs to be lightly doped to maintain a small junction capacitance. This in turn produces a relatively large value of intrinsic base resistance. If the supply of impurity material is limited then the depth of alloying is dependent on the temperature and the wetted area. When the impurity is in the form of a wire, rather than a button, then the impurity is unlimited, and the time of alloying must also be controlled.

The alloying process is repeated on the other side of the crystal to obtain the required two-junction structure, such as Fig. 3.3.

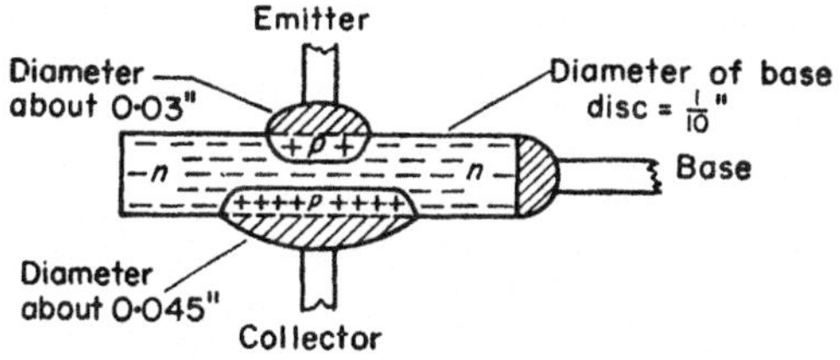

FIG. 3.3. Alloyed *pnp* general purpose transistor, with typical dimensions.

Contacts are soldered to the *n* type base region and to the emitter and collector pellets. To minimize the base resistance the base contact is preferably in the form of a ring round the emitter.

It is difficult to alloy into a wafer of less than 0·05 mm thickness (say, 0·002 in.) and extremely difficult to make alloyed transistors with base widths more narrow than 0·025 mm consistently. This results in a usable upper frequency limit of about 10 Mc/s.

Alloying, and in particular the requirement to maintain a small junction capacitance, is dealt with in more detail in some of the references given in Appendix B.

3.4. Grown Junctions

Two main methods may be used to grow a single crystal with alternate *p* and *n* layers.

The double doping method grows a crystal, as in Fig. 3.1, except that when a suitable length of *n* type material has been grown, then trivalent material is added to the melt. This must be in sufficient quantity to swamp the pentavalent material already present. *p* Type material is then pulled out of the melt for a short time, and then more pentavalent material is added to produce a layer of *n* type material.

An alternative is known as Rate Growing. An *n* type crystal is pulled from a melt having both *p* and *n* impurities. Both these

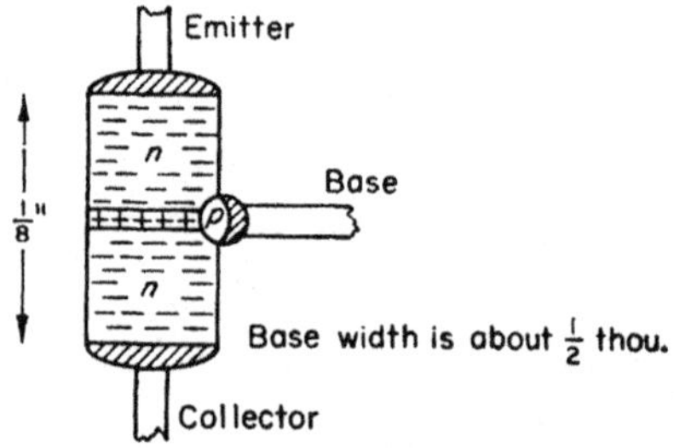

FIG. 3.4. Grown junction *npn* transistor.

impurities are fairly concentrated, but there is an effective preponderance of *n* type. However, a transient reduction in the growth rate reduces the *n* type concentration more than the *p* type, giving a *p* layer sandwiched between two *n* layers. By this process several *npn* arrays may be produced in one crystal.

From a crystal prepared by one of these methods are cut bars, about 1 mm square and 3 mm long, containing the junctions. These are always in the form of *npn* transistors. Emitter and collector leads are soldered to opposite ends of the bar. The main manufacturing problem with a grown junction transistor is to make a satisfactory contact with the narrow base region. Usually, a base lead is connected by placing a fine wire in contact with the base region and alloying the wire into the base with a carefully controlled current pulse. In the case of germanium a gold wire, with a small proportion of gallium, is used. Although the wire will often dissolve some *n* type germanium, as well as some of the *p* type base, this is of no significance as the presence of the gallium produces recrystallized *p* type germanium. The end of the base contact then appears as a small bulge in the base layer, as shown in Fig. 3.4.

A base width of about 0·0015 mm, or 1·5 micron (μ) may be achieved in a grown junction transistor, with an associated collector capacitance of 3 pF. This compares with a minimum collector junction capacitance in an alloyed junction transistor of about 10 pF. On the other hand, a disadvantage of the grown transistor is the relatively high internal series resistance between each connection and the actual junctions.

3.5. Diffusion

A region of semiconductor material may be converted from one type to another by solid diffusion. This process gives excellent control of impurity penetration to within a few millionths of an inch, and is therefore attractive as a way of producing the thin base region needed for a successful very high frequency transistor.

If a *p* type germanium wafer is heated to near melting in a donor impurity atmosphere, such as antimony, then some of the impurity condenses on the wafer surface and tends to diffuse into the crystal. When the wafer is cool it may be proved that the thin surface layer of germanium has changed from a *p* to an *n* type semiconductor. The position of the junction within the germanium and the gradient of the impurity concentration are determined by the concentration of impurity in the original semiconductor and the product of the time and the Diffusion Coefficient of the impurity being diffused. In fact this diffusion coefficient is a function of the nature of the impurity and the temperature. Since the depth of junction can be made very small with only moderate control of diffusion coefficient and time, this is a very powerful method.

A *pnp* structure may be formed by alloying into the *n* type layer with, for example, aluminium which has been evaporated through a mask on to the surface of the germanium wafer. A contact to the base region is made with an evaporated gold–antimony alloy. An etching process reduces the collector junction area, and produces a transistor similar in cross-section to that of Fig. 3.5. Since both the base and emitter appear as a small plateau, this is generally known as the mesa transistor.

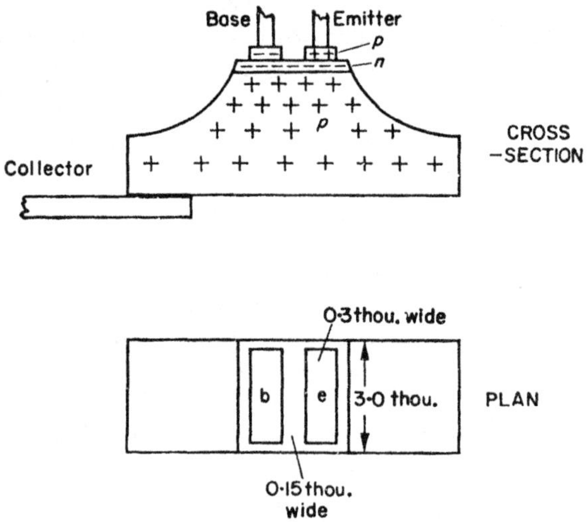

FIG. 3.5. Diffused mesa transistor.

It is a significant fact, but not thoroughly understood, that *n* type impurities diffuse fairly rapidly into germanium, but that the diffusion rate of *p* impurity into germanium is so slow as to make it an unusable process. But in silicon diffusion of both *n* and *p* impurity is a worth-while proposition, to give a double-diffused *npn* transistor. Using this technique it is possible to produce a transistor with a base width of 0·5 μ.

The technique of manufacture of double-diffused silicon transistors depends on the selective masking property of silicon dioxide. A layer of this dioxide, about 0·5 μ thick, has little effect upon the diffusion of gallium into the silicon from a gallium-rich atmosphere. But below 1300°C this same dioxide layer effectively masks the diffusion of boron, phosphorus, arsenic or antimony. Using this property of silicon dioxide, together with refined photo-lithographic techniques, a high performance double-diffused transistor is made by the following steps:

(a) Preparation of a flat very smooth silicon slice of *n* type material.
(b) Growth of a silicon dioxide layer, between 0·5 and 1 μ thick.
(c) Diffusion of gallium through dioxide to give *p* base layer.
(d) Emitter masking by the photo-lithographic process, making a pattern of small holes in the dioxide layer, through which an *n* type element will be diffused. This gives a small emitter and low input capacitance.
(e) Diffusion of phosphorus to form the emitters.
(f) Mesa masking by another photo-lithographic technique, followed by etching. By limiting the area of the collector–base junction this gives a low output capacitance.
(g) Evaporation of aluminium through a metal mask to form base and emitter contacts.
(h) Breaking up the slice into as many as one thousand separate transistor units and mounting in suitable encapsulation.

3.6. Epitaxial and Planar Techniques

The diffused mesa transistor has several non-ideal characteristics, of which the most significant are high saturation resistance (which results in a troublesome value of knee voltage), low gain at small values of current and poor noise figure. In a double-diffused silicon transistor the active part of the device occupies less than 10 μ of the slice thickness. But the minimum slice thickness for satisfactory handling in mass production is about 100 μ (or 0·004 in.). In effect the bulk of the collector material provides mechanical support, whilst having relatively high electrical and thermal resistivity. Or, in other words, the construction of a semiconductor structure with a narrow base width does not necessarily make a good transistor.

The chemical process known as epitaxy has been revived to deposit single crystal layers of semiconductor on a single crystal substrate. For instance, a thin layer (10 μ) of high resistivity (5 Ω-cm) silicon is deposited epitaxially on a very low resistivity (0·001 Ω-cm) substrate, of about 100 μ thickness. The transistor may then be fabricated in the upper 10 μ layer in the manner described in Section 3.5.

Two main ways of obtaining the thin epitaxial layer of high resistivity are available. The more popular involves the reduction of germanium or silicon chloride with hydrogen at high temperatures of 800° or 1200°C respectively. The tetrachloride gas ($GeCl_4$ or $SiCl_4$) is passed over the germanium or silicon substrate and a deposition rate of 1 μ/min is usual.

Generally, the epitaxial layer is of the same type of semiconductor as the substrate (i.e. n on n^+ or p on p^+), but it is possible to grow either n or p type layers. Diodes may thus be formed without any diffusion steps, and transistors may eventually be made in this way.

The planar transistor differs from the mesa type in that the geometry of the collector–base junction is defined by oxide masking and photo-lithographic technique, and not by chemically etching away the corners to form the mesa. An oxide layer is grown above the collector wafer, which has been formed by the epitaxial process as described above. A pattern of holes is then etched, in this oxide layer, of the same geometry as that in the mesa type of transistor. Boron is then diffused through these holes into the silicon, and actually diffuses laterally about the same distance as it diffuses normally to the surface. So the collector–base junction is really produced underneath the inert silicon dioxide layer. Then the emitter is produced by diffusion, using oxide grown during the boron diffusion for masking purposes, and the emitter–base junction is also under the oxide layer. Small holes are carefully etched in the oxide, at appropriate points, to allow very thin wires to be pressure welded to the base and emitter regions. Since the sealed junctions give very stable characteristics the planar transistor has several advantages, such as:

(a) Very low leakage currents, of the order of 0·1 nA (10^{-10} A) in a small-signal device.
(b) Enhanced reliability as the junctions are not exposed to contaminating chemicals during manufacture and are sealed when the transistor is being encapsulated.
(c) High current gain at low currents, and low noise figures.

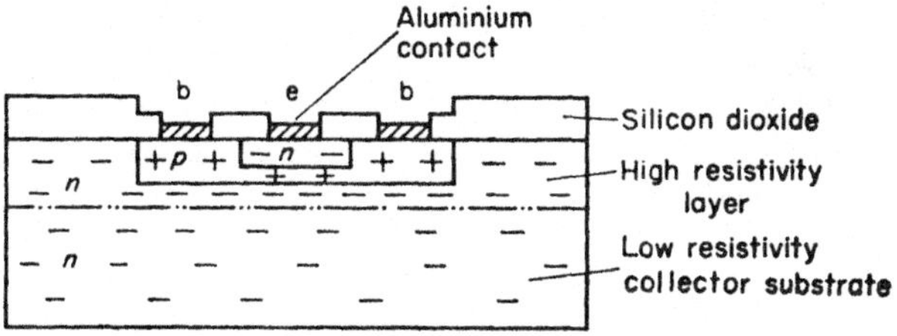

FIG. 3.6. Epitaxial planar silicon transistor.

A cross-section of a typical epitaxial planar silicon transistor is given in Fig. 3.6.

3.7. Encapsulation

However the active part of a transistor is made, it has to be put into an enclosure to protect from mechanical and chemical damage. An all-glass encapsulation follows the technique developed for valves, often with the transistor coated in silicone grease. This gives a fragile envelope, not suitable for operation at higher powers or temperatures.

A metal enclosure is now most popular, usually consisting of a base carrying the electrode leads and a cover, to which the transistor is sometimes directly connected. Several methods of joining the two metal parts are available, the main object being to avoid the use of excessive heating. It seems likely that a transistor made by the planar epitaxial technique would be less likely to suffer contamination during encapsulation than devices made by other processes.

In transistors designed to operate at ultra-high frequencies the inductance of the leads and self-capacitance within the can are of great significance. These effects often degenerate the performance of a semiconductor structure which may, in itself, have a good current gain in the U.H.F. range, and mask the measurement of the actual transistor parameters. These U.H.F. transistors have to be considered as complete assemblies including envelopes and mountings, for, although the actual transistor may be very good before encapsulation this may be ruined by stray inductance and capacitance.

3.8. D.C. Characteristics

In this section we shall consider the static curves of a transistor for the cases where either the emitter or the base connection is common (or grounded). The characteristics for common collector are not usually discussed separately. For this purpose it is general to use the grounded base or grounded emitter knowledge, suitably modified to allow for the different conditions. As an example the appropriate values will be given for a low power *pnp* transistor intended for general purpose use.

The two important curves are the input and output characteristics. The first of these is a current/voltage curve, looking like that of a forward biased diode. Using the current and voltage directions as defined in Fig. 3.7, a set of common base input curves are given in Fig. 3.8. The output curves in Fig. 3.9 look very much like a series of diode characteristics, with reverse bias, which are displaced along the current axis in steps (at practically equal intervals). At voltages near the zener value the curves turn sharply upward, because of avalanche breakdown in the collector junction.

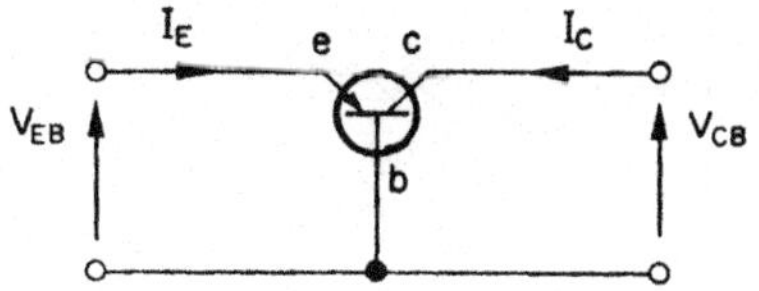

FIG. 3.7. Currents and voltages in common base.

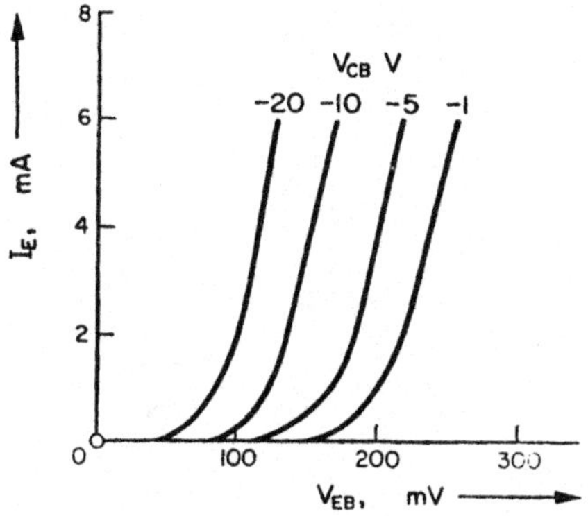

FIG. 3.8. Common base input curves.

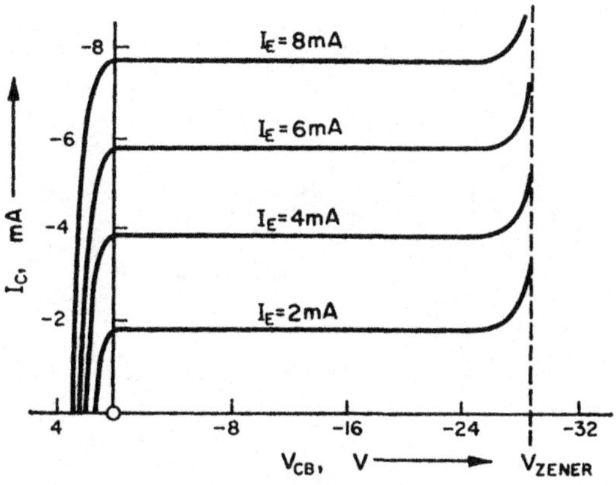

FIG. 3.9. Common base output curves.

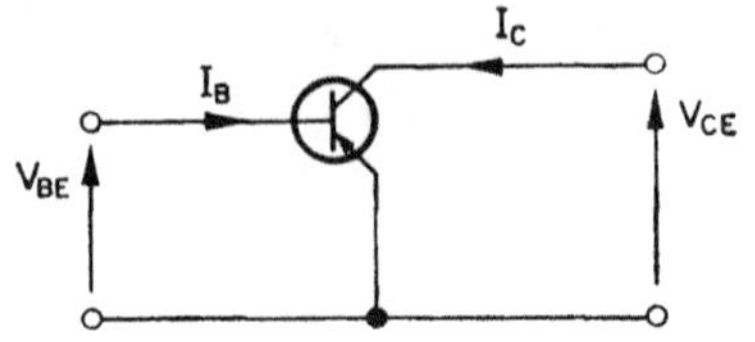

FIG. 3.10. Currents and voltages in common emitter.

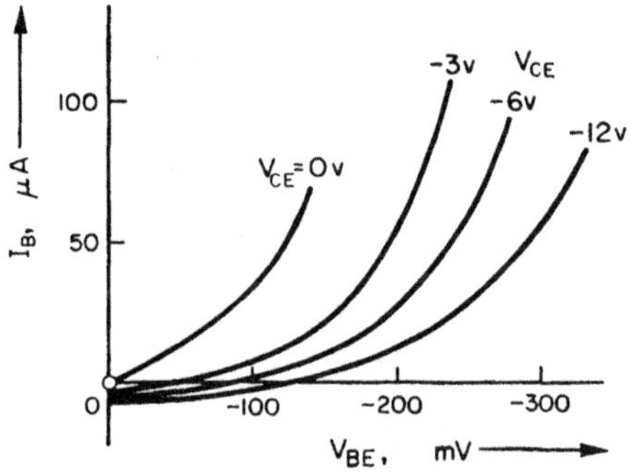

FIG. 3.11. Common emitter input curves.

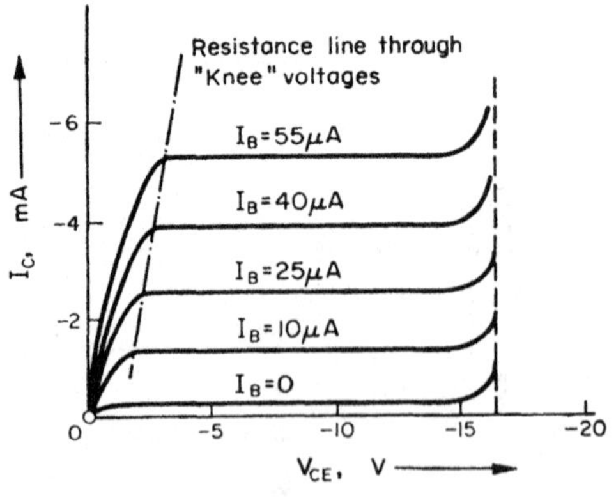

FIG. 3.12. Common emitter output curves (showing " knee " voltages).

Note that the input resistance depends on V_{EB}, but is small for values of V_{EB} greater, in this case, than about 150 mV. Also the output resistance is generally large (between the " knee " value and the approach of breakdown).

The current and voltage directions for the common emitter curves are defined in Fig. 3.10. The input curves of Fig. 3.11 show that the input resistance is higher with common emitter than it is for the common base connection. Also in Fig. 3.12 there is a significant value of collector current I_c for the condition when the base current I_B is zero. This is known as " leakage current ", and for the common emitter mode is generally denoted by the symbol I'_{co}. To reduce the collector current to zero when a collector to emitter voltage is applied requires that the base current be reversed. Normally, the best that one can do in a circuit is to reduce base current to zero, but some switching circuits do make provision for a reversal of base current.

3.9. Leakage Currents

The magnitude of leakage current within a transistor is temperature dependent, as is also the base to emitter voltage. In the case of germanium the leakage current approximately doubles for a 7°C rise from 20°C and in silicon the leakage current doubles for a 10°C rise from 30°C (but is normally very small in a silicon device). These factors mean that the inherent power dissipation with a germanium transistor increases in a significant manner as the junction temperature rises. So at a critical junction temperature the " thermal feedback loop " within the transistor has a gain which is greater than unity, and " thermal runaway " occurs. That is to say that, under certain conditions, the internal dissipation in the transistor, due to leakage current, may have a cumulative heating effect. For any transistor, however small the leakage current may actually be at room temperature, there will be a junction temperature at which thermal instability results. This leads to the circuit requirement of stabilization of the operating point.

To achieve transistor operation at high temperatures and high power it is necessary to have low initial leakage current and low thermal impedance away from the collector junction. The silicon transistor achieves the first of these requirements, and a large collector area connected directly to the metal casing of a transistor helps to achieve the second.

3.10. Current Gain

Since the input resistance of a transistor (in either common base or common emitter mode) changes with the voltage applied, as shown by the non-linear input curves, so it is usual to drive the

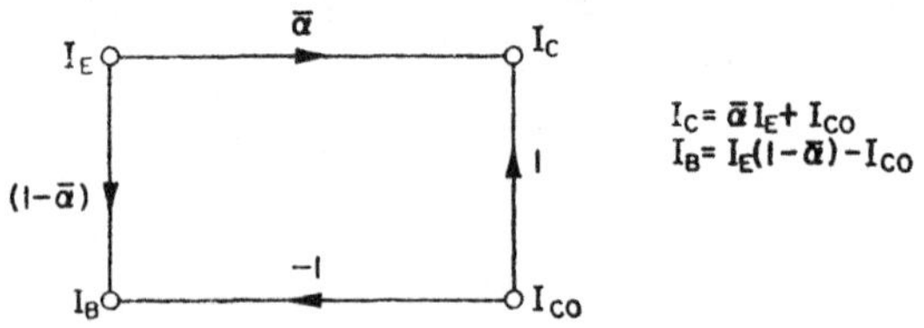

FIG. 3.13. Flow diagram of direct current in a transistor.

transistor and bias it from current sources. This means that the effective source resistance for both signal and bias must be large in comparison with the input resistance of the transistor. If the constant impedance of the source is not high enough to swamp the varying input impedance of the transistor, then a large amount of distortion will result at the input circuit.

With current drive, it is clear that the current gain of the transistor is an important property. For slowly varying signals the current gain may be found from the static or dynamic curves drawn for the d.c. conditions. These values are greatly affected by the frequency of the signal and the presence of low or high values of current. The simple definitions for low frequency current gain in common base and common emitter connections are

$$\alpha_0 = \left.\frac{\partial I_c}{\partial I_E}\right| V_{CB} \text{ constant} \quad \text{and} \quad \beta_0 = \left.\frac{\partial I_c}{\partial I_B}\right| V_{CE} \text{ constant.}$$

The large signal values of current gain involve large changes of current and we then consider the transistor to be Off when $I_E = 0$. This definition takes into account the leakage current which may flow when the emitter lead is open-circuited. So the current gains for large signals in common base and common emitter are

$$\bar{\alpha} = \frac{\Delta I_c}{\Delta I_E} = \frac{I_c - I_{co}}{I_E} = \frac{I_c - I_{co}}{I_c + I_B}$$

and

$$\bar{\beta} = \frac{\Delta I_c}{\Delta I_B} = \frac{I_c - I_{co}}{I_B + I_{co}}.$$

These quantities may be represented in the flow diagram, having the three nodes I_E, I_B and I_c, of Fig. 3.13. Very often the appropriate signs for the collector and base currents are ignored in the published data, and this has been done in Fig. 3.13.

Since

$$\alpha_0 = \frac{\partial I_c}{\partial I_E} \quad \text{and} \quad \beta_0 = \frac{\partial I_c}{\partial I_B}, \quad \text{so} \quad \frac{\alpha_0}{\beta_0} = \frac{\partial I_B}{\partial I_E}.$$

Also, $\quad I_B = I_E - I_c, \quad \text{so} \quad \partial I_B = \partial I_E - \partial I_c.$

Therefore

$$\frac{\alpha_0}{\beta_0} = 1 - \frac{\partial I_c}{\partial I_E} = 1 - \alpha_0$$

or $$\beta_0 = \frac{\alpha_0}{1-\alpha_0}.$$

When the base of a transistor is open-circuited ($I_B = 0$), then the leakage current is $I'_{co} = I_E = I_c$. Or if the emitter is open-circuited, then $I_B = I_c = I_{co}$. From the flow diagram of Fig. 3.13 we can see that $(1-\bar{\alpha})I_E = I_{co}$, if the base is, in fact, open-circuited. Combining these expressions leads to the statement that $I'_{co} = I_E = I_{co}/(1-\bar{\alpha})$ for open-circuit base. This means that the leakage current flowing through the transistor is much larger in common emitter than with the common base connection.

Due to the significant transit time for the majority charge carriers (holes in a *pnp* transistor) from emitter to collector, the current gain of a transistor falls with increasing signal frequency. In the common emitter connection it is usual to define either a frequency (f_{co}) at which current gain has fallen by -3 db, or a frequency (f_1) at which the current gain has fallen to unity, such that $i_c = i_b$. Due to the impedance transformation through a transistor some voltage gain is obtainable even at the frequency f_1.

3.11. Power Dissipation

The base to emitter voltage within a transistor is always small (often less than 0·5 V) and the emitter junction dissipation is correspondingly small. So the total transistor power dissipation may be considered as approximately the same as the collector dissipation ($P_c W$). The maximum permissible junction temperature ($T_{j(\max)}$) is often taken as 80°C for germanium and 170°C for silicon. Other semiconductor materials may give higher permissible junction temperatures.

The maximum possible collector dissipation ($P_{c(\max)}$) depends on the thermal impedance between the collector-to-base junction and ambient conditions. It is usual to consider ambient conditions as an infinite heat sink, but this assumption must be treated with caution and does not apply in certain equipment designs. If this thermal impedance from collector junction to ambient is given by θ°C/W and if the ambient temperature is T_{amb}°C, then the maximum collector dissipation is

$$P_{c(\max)} = \frac{T_{j(\max)} - T_{amb}}{\theta} \text{ W}.$$

A wide range of values for thermal impedance exists in practice. For a low power transistor in a glass envelope without a heat sink, a typical value is $\theta = 0{\cdot}5$°C/mW (or 500°C/W), whilst for a medium power device with a heat sink $\theta = 10$°C/W may be typical, and a value of $\theta = 1$°C/W will only be achieved by using a large power transistor with some degree of forced cooling.

Conduction, convection and radiation all assist in the removal of heat from the transistor, so that an exact thermal circuit would be difficult to formulate. We may consider the total thermal resistance from junction to ambient as made up from several thermal resistances in series.

The resistance from collector junction to mounting base (θ_m) is dependent on the construction of the transistor. Of the four main types described earlier in this chapter, the epitaxial transistor should have the least value of θ_m, and the alloyed junction type will have the highest thermal resistance. The resistance from the transistor can to the heat sink, which will often be a metal chassis, depends on whether there is direct metallic contact or some electrical insulation, such as a mica washer, is included. This is given the symbol θ_i°C/W. Finally, the resistance from heat sink to free air (θ_h) depends on the shape, area and material of the heat sink and on the velocity of air flow. It therefore follows that the maximum collector power dissipation may be calculated from

$$P_{c(\text{max})} = \frac{T_{j(\text{max})} - T_{\text{amb}}}{\theta_m + \theta_i + \theta_h}.$$

Clearly, the maximum permissible junction dissipation decreases with an increase in ambient temperature and also depends to a large extent on the nature of the heat sink used.

For example, assume a germanium power transistor, with mica insulation to the chassis, has a collector dissipation of 5 W and the maximum expected ambient temperature is 45°C. Also the heat sink has $\theta_h = 3$°C/W, the insulation makes $\theta_i = 0{\cdot}5$°C/W and the transistor thermal resistance is $\theta_m = 4$°C/W.

Then the heat sink temperature is $= 45° + 5 \times 3° = 60$°C.

Therefore the mounting base temperature

$$= 60° + 5 \times 0{\cdot}5° = 62{\cdot}5°\text{C}.$$

This results in a junction temperature $= 62{\cdot}5° + 5 \times 4° = 82{\cdot}5$°C.

Any design can be checked for thermal stability in this way, providing that the thermal resistances can be found. The thermal resistance of any heat sink arrangement may be calculated by measuring the ambient temperature and the heat sink or mounting base temperature.

For instance,

$$\theta_h = \frac{T_{mb} - T_{\text{amb}}}{P_c} - \theta_i = \frac{T_{hs} - T_{\text{amb}}}{P_c} \quad (^\circ\text{C/W}).$$

Questions for Chapter 3

Q.3.1. Give sketches of the d.c. characteristics of a *pnp* junction transistor, suitable for general purpose low frequency applications, in the common emitter configuration.

Show how the general shape of the curves is accounted for by the physical mechanism of operation. Sketch a simple equivalent circuit to represent this transistor under small signal conditions and indicate how the parameters of the circuit are derived from the characteristic curves.

Q.3.2. What are the relationships between the current gains and the leakage currents of a transistor in common base and common emitter connections? Explain the importance of these differences, with particular reference to operation at high frequency or high temperature.

Q.3.3. Sketch labelled diagrams showing a cross-section of the semiconductor material and the connecting leads for the following types of transistor:

Alloy Junction,
Grown Junction,
Alloy-diffused Mesa,
Planar Epitaxial.

Give a typical dimension for the base width in each case. State briefly the properties required for successful operation with worth-while gain at high frequencies, and compare the four transistors on this basis.

Q.3.4. A power transistor, in an amplifier, dissipates 4 W. Calculate the maximum permissible value of the thermal resistance from heat sink to free air, if the mounting base and contact have a total thermal resistance $(\theta_m + \theta_i)$ of 4°C/W, under the following conditions:

(a) Germanium transistor with $T_{j(\max)}$ = 85°C.

(b) Silicon transistor with $T_{j(\max)}$ = 170°C.

Assume the maximum expected ambient temperature is 100°F in each case.

CHAPTER 4

Equivalent Circuits

ALTHOUGH many transistor equivalent circuits have been suggested and used, the equivalent " T " bears the closest relation to the physical construction of the transistor. It will be used to analyse the operation of typical circuits in later chapters and is emphasized in this, but reference is also made to some of the other commonly used circuits. Throughout the chapter small letters will be used for a.c. values, while d.c. values are shown in capitals.

4.1. Common Base, Low Frequency, Equivalent *T* Circuit

With correct operating potentials, a transistor comprises a forward biased emitter junction, between a low resistivity emitter and a relatively high resistivity base, and a reverse biased collector junction between the base and low resistivity collector. The resistances of the emitter and collector regions are very low but account must be taken of the appreciable base resistance, and this may be shown as a resistance $r_{bb'}$, where b is the external base connection and b' an imaginary internal base connection. Figure 4.1 shows the complete low frequency equivalent circuit with r_e as the resistance of the forward biased emitter diode and r_c the resistance of the reverse biased collector diode. Current and voltage generators complete the circuit, the current generator $\alpha_0 i_e$ representing the proportion of the emitter current i_e which appears at the collector, while the voltage generator $\mu v_{cb'}$ represents feedback between the collector and emitter circuits. The latter is necessary since an increase in collector voltage leads to an increase in depletion layer thickness and decrease in base width, which implies a lower emitter current for a given collector current. The factor μ, however, is not independent since it is related to r_e, r_c and α_0 by the relation $\mu = r_e/2r_c(1-\alpha_0)$. This equation is proved in Appendix C.

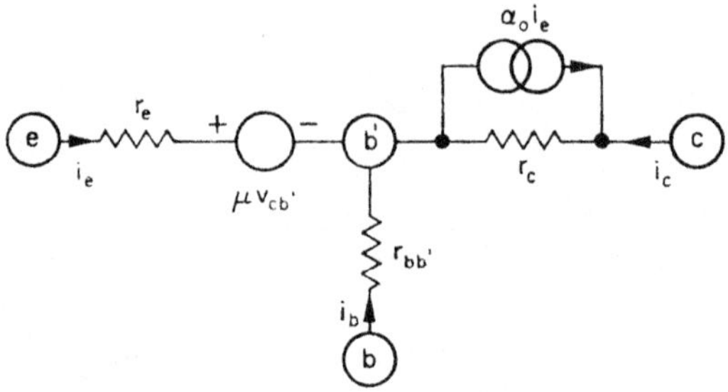

FIG. 4.1. Grounded base equivalent circuit.

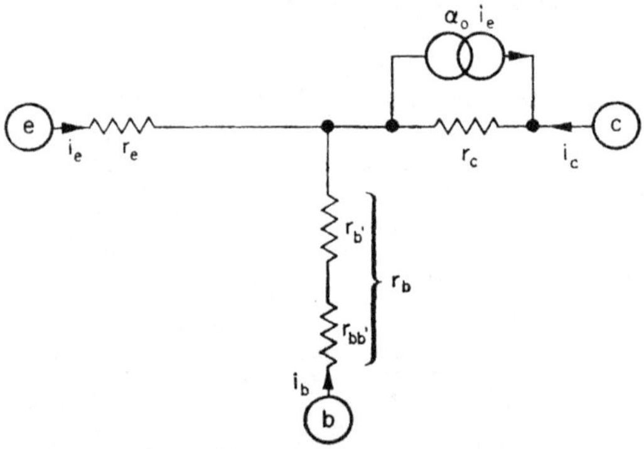

FIG. 4.2. Simplified grounded base equivalent circuit.

The equation for the input current of Fig. 4.1 is

$$v_{eb} = i_e r_e + \mu v_{cb'} + (i_e + i_c) r_{bb'} = i_e r_e + \mu(i_c + \alpha_0 i_e) r_c + (i_e + i_c) r_{bb'}$$

$$\simeq i_e r_e + (i_c + i_e)(\mu r_c + r_{bb'}), \quad \text{since} \quad \alpha_0 \simeq 1.$$

Hence for low frequency purposes $\mu v_{cb'}$ may be replaced by an additional resistance $r_{b'}$ in the base lead as shown in Fig. 4.2. The value of $r_{b'}$ is μr_c or $r_e/2(1-\alpha_0)$ and this is known as the intrinsic base resistance whereas $r_{bb'}$ is known as the extrinsic base resistance.

The parameters in the equivalent circuit are dependent on the operating potentials and currents and r_e and $r_{b'}$ are inversely proportional to the standing emitter current I_e, while r_c is approximately inversely proportional to I_c, the standing collector current. The short-circuit current gain α_0 is dependent on both I_e and V_c, the standing collector voltage, and variation of α_0 with I_e is shown in Fig. 4.3. Initially, as I_e is increased the carriers drawn into the

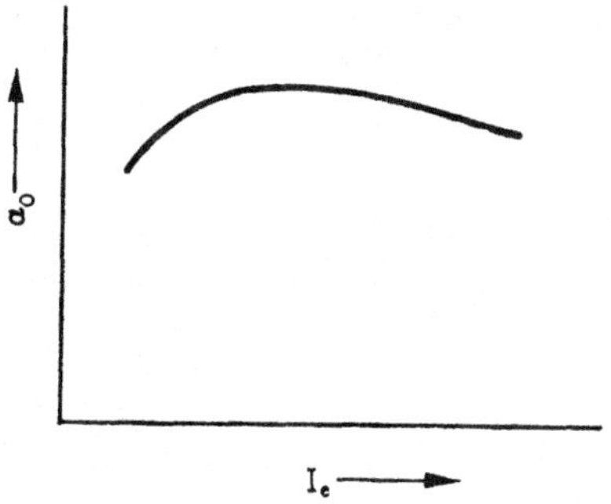

FIG. 4.3. Variation of current gain with emitter current.

base to maintain charge neutrality develop their own field which assists the flow of carriers from the emitter, but as I_e is increased further the number of carriers drawn into the base becomes comparable with the minority carrier density in the base, and the emitter efficiency is reduced. The effect of collector voltage V_c on α_0 is to increase its value as the breakdown voltage $V_{c'}$ is approached. This is an avalanche effect and its result may be given mathematically by dividing the low voltage value of α_0 by $[1-((V_c)^n/V_{c'})]$, n being about 3 for an alloy junction transistor.

The low frequency circuit shown in Fig. 4.2 is sometimes converted into the form shown in Fig. 4.4 by the application of Thevenins Theorem. Figure 4.5 shows the equivalence, and equating the open-circuit voltage for each case gives:

$$\alpha_0 i_e r_c = r_m i_e \quad \text{or} \quad r_m = \alpha_0 r_c.$$

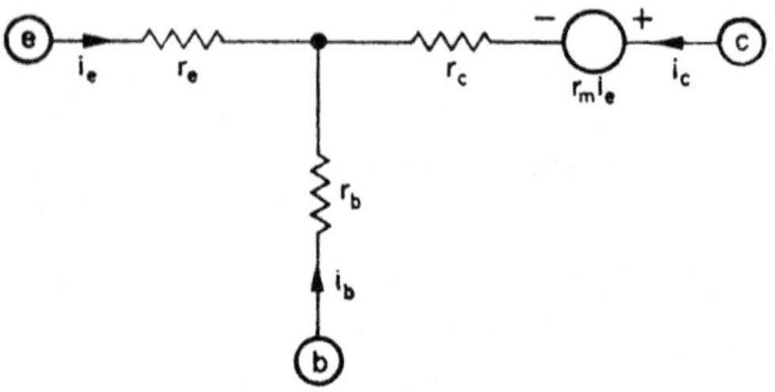

FIG. 4.4. Simplified grounded base equivalent circuit, with voltage generator.

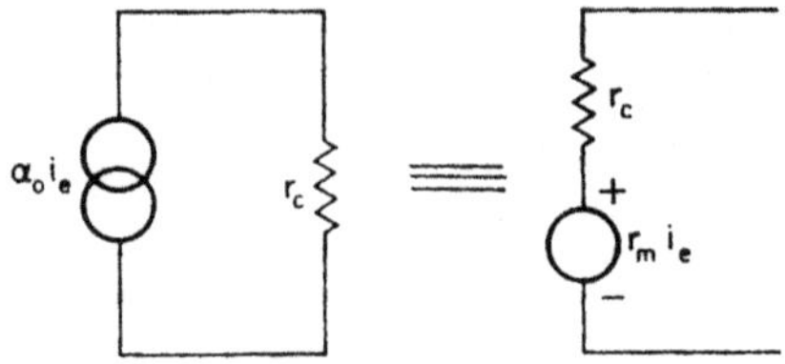

FIG. 4.5. Relation between current and voltage generators.

4.2. Common Base, High Frequency, Equivalent *T* Circuit

As the operating frequency is increased emitter and collector capacitances c_e and c_c must be considered. The former is composed of the junction capacitance and the much larger diffusion capacitance, but in the latter case only the junction capacitance is important. The diffusion capacitance is due to carriers diffusing through the base and is a measure of the charge stored in the base. These considerations lead to the high frequency equivalent circuit shown in Fig. 4.6. Normally, for high frequency circuits, the reactance of c_c is very much less than r_c and the alternating voltage appearing at the collector is low enough to ignore $\mu v_{cb'}$. Also the current i_1, which allows for the part of the emitter current

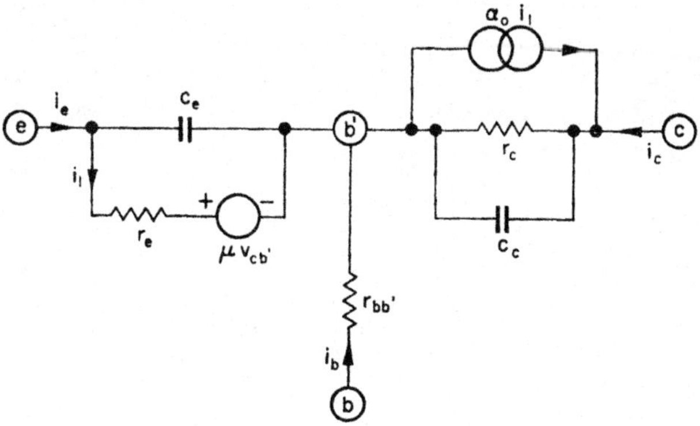

FIG. 4.6. High frequency, grounded base, equivalent circuit.

flowing through the emitter capacitance is related to the total emitter current i_e by the relation

$$i_1 = \frac{i_e}{1+j\omega c_e r_e}$$

and this gives an expression for the high frequency current gain:

$$\alpha_0 i_1 = \frac{\alpha_0 i_e}{1+j\omega c_e r_e}$$

$$= \alpha i_e, \quad \text{where} \quad \alpha = \alpha_0/1+j\omega c_e r_e.$$

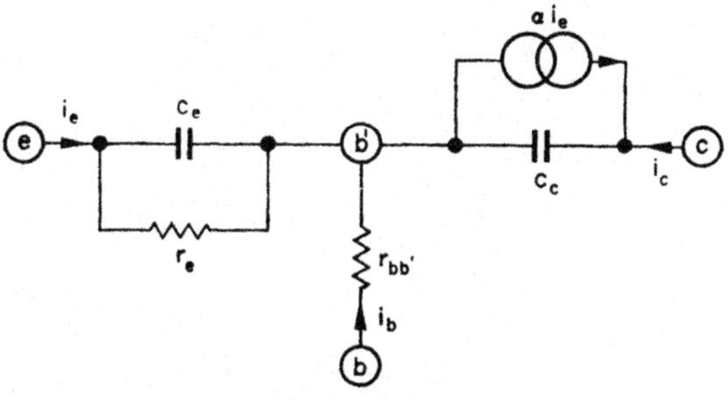

FIG. 4.7. Simplified, high frequency, grounded base equivalent circuit.

The high frequency circuit may now be redrawn as shown in Fig. 4.7.

The reduction of current gain at high frequencies is a transit time effect and is important as the time of travel of carriers across the base becomes comparable with the period. Then

$$\alpha = \frac{\alpha_0}{1+j\omega c_e r_e}$$

$$= \frac{\alpha_0}{1+j\omega/\omega_\alpha}, \quad \text{where} \quad \omega_\alpha = 1/c_e r_e,$$

$$= \frac{\alpha_0}{1+jf/f_\alpha},$$

where f_α is defined as the frequency where the current gain falls to $1/\sqrt{2}$ of the low frequency value.

4.3. Common Emitter, Low Frequency, Equivalent *T* Circuit

This may be obtained by re-orientating the grounded base equivalent *T* circuit; Figs. 4.8, 4.9 and 4.10 showing different variations of the common emitter equivalent *T*. It is often more convenient to give the current generator shown in Fig. 4.9 in terms of the alternating base current i_b and the transformation is easily made by considering the voltage drop across r_c,
i.e. voltage drop across r_c

$$= i_c r_c + \alpha_0 i_e r_c$$

$$= i_c r_c - \alpha_0 (i_c + i_b) r_c, \quad \text{since } i_e + i_b + i_c = 0,$$

$$= r_c(1-\alpha_0) i_c - \alpha_0 i_b r_c$$

$$= r_c(1-\alpha_0)\left[i_c - \left(\frac{\alpha_0}{1-\alpha_0}\right) i_b\right]$$

$$= r_c(1-\alpha_0)\,[i_c - \beta_0 i_b], \quad \text{where } \beta_0 = \alpha_0/(1-\alpha_0).$$

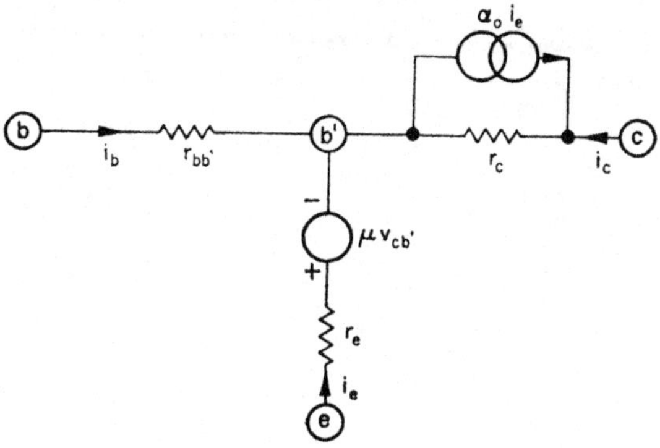

FIG. 4.8. Grounded emitter equivalent circuit.

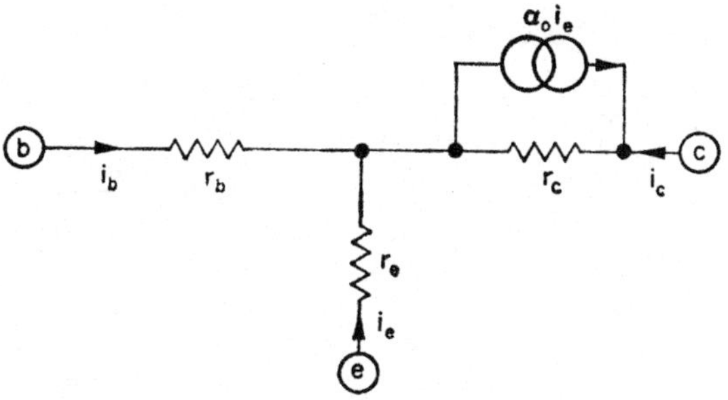

FIG. 4.9. Simplified grounded emitter equivalent circuit.

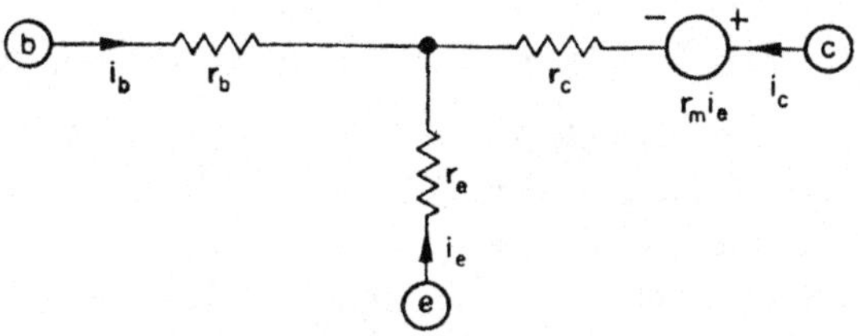

FIG. 4.10. Simplified grounded emitter equivalent circuit with voltage generator.

Figure 4.9 may now be transformed into Fig. 4.11 which gives the current generator in terms of i_b, but a lower value of resistance in the collector circuit.

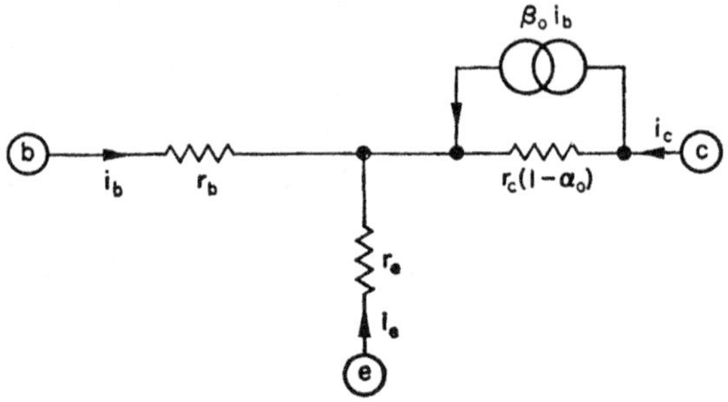

FIG. 4.11. Modified grounded emitter equivalent circuit.

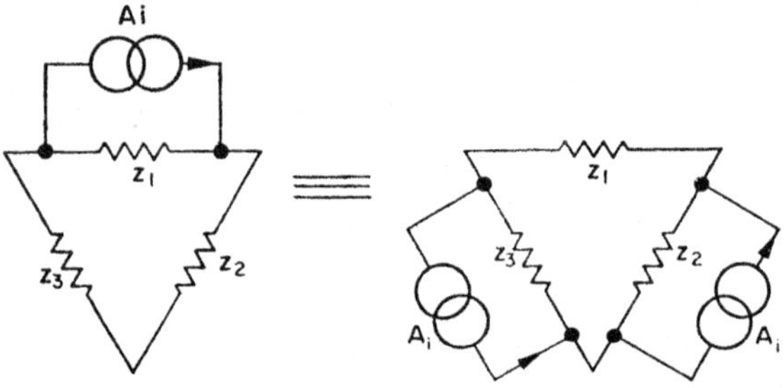

FIG. 4.12. Current generator—theorem.

4.4. Common Emitter, Hybrid π Equivalent Circuit

The hybrid π equivalent circuit is often used for the common emitter configuration and may be derived from the equivalent T by means of the theorem shown in Fig. 4.12. The theorem shows that a current generator across one of the three impedances connected in delta may be replaced by identical current generators across the other two, and may easily be verified by calculating the voltage across any impedance for both configurations.

Figure 4.8 shows that $i_e = (v_{eb'} - \mu v_{cb'})/r_e$ and this current may be substituted in the current generator $\alpha_0 i_e$. Also $\mu v_{cb'}$ in series with r_e may be replaced by the equivalent current generator to

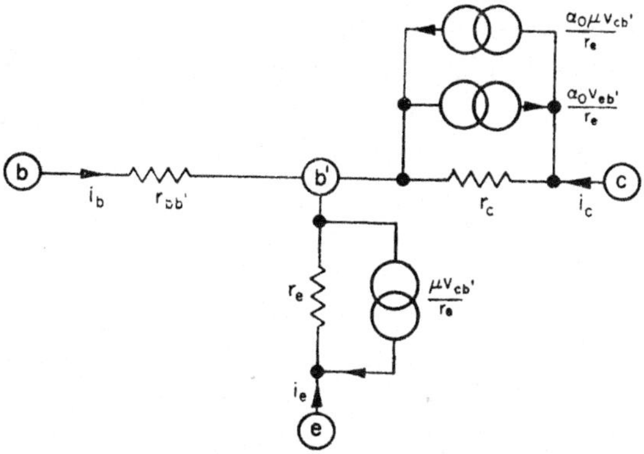

FIG. 4.13. Development of hybrid π equivalent circuit—stage 1.

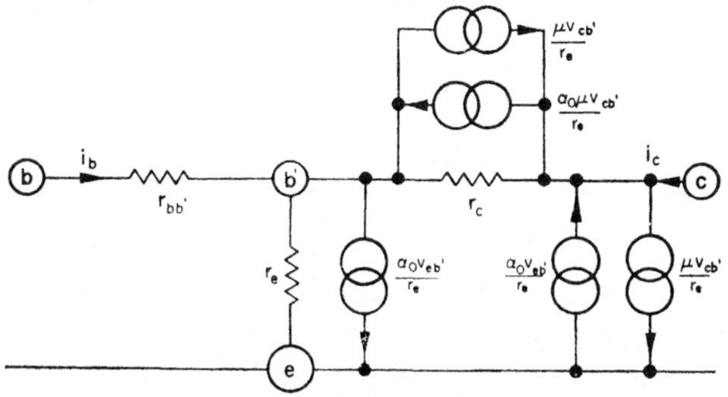

FIG. 4.14. Development of hybrid π equivalent circuit—stage 2.

give Fig. 4.13. The current generators $\mu v_{cb'}/r_e$ and $\alpha_0 v_{cb'}/r_e$ may each be replaced by two others to give the equivalent circuit shown in Fig. 4.14 and further reduced by converting some of the generators to resistors.

For the link eb', $\alpha_0 v_{eb'}/r_e$ may be replaced by a resistance $-r_e/\alpha_0$ which in parallel with r_e gives

$$r_e\left[-\left(\frac{r_e}{\alpha_0}\right)\Big/\left(r_e-\frac{r_e}{\alpha_0}\right)\right]=\frac{r_e}{1-\alpha_0}$$

denoted by $r_{b'e}$. For cb' the current generators may be combined to give a single generator $\mu(1-\alpha_0)v_{cb'}/r_e$ flowing from b' to c, and

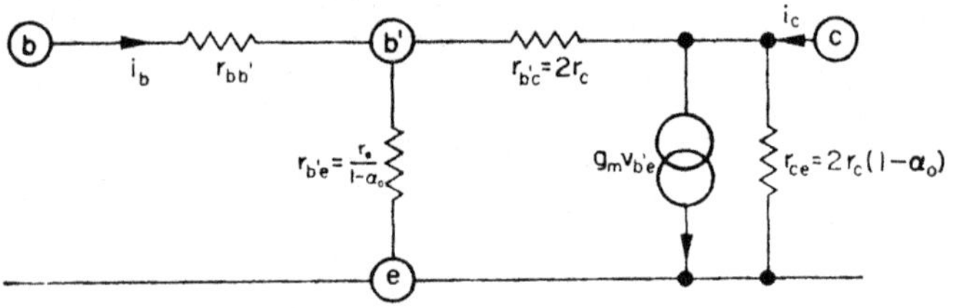

FIG. 4.15. Hybrid π equivalent circuit.

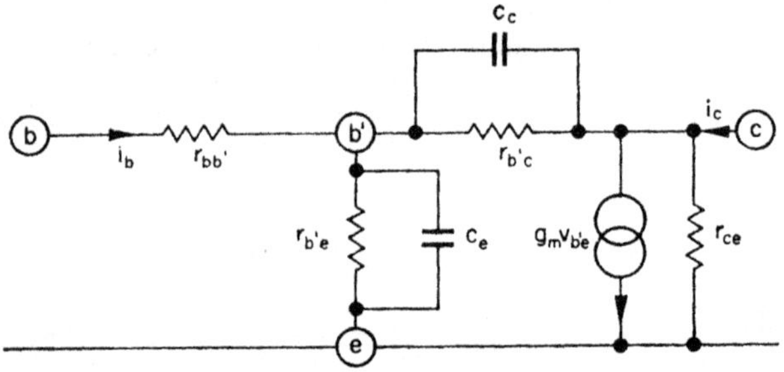

FIG. 4.16. High frequency hybrid π equivalent circuit.

since $\mu = r_e/2r_c(1-\alpha_0)$ this may be replaced by a current generator $v_{cb'}/2r_c$, and hence by a resistor $-2r_c$ which in parallel with r_c gives $2r_c = r_{b'c}$. In practice r_e is low and $v_{cb'}$ is approximately equal to v_{ce} so that the current generator $\mu v_{cb'}/r_e$ may be replaced by a resistance $r_e/\mu = 2r_c(1-\alpha_0) = r_{ce}$. Using these substitutions we may derive the hybrid π equivalent circuit shown in Fig. 4.15 where $g_m = \alpha_0/r_e$.

This form of the grounded emitter equivalent circuit is particularly useful at high frequencies when c_e may be shown across $r_{b'e}$ and c_c across $r_{b'c}$ as in Fig. 4.16.

4.5. Common Collector Equivalent *T* Circuit

This again may be obtained by re-orientating the common base circuit, Fig. 4.2 being modified to Fig. 4.17.

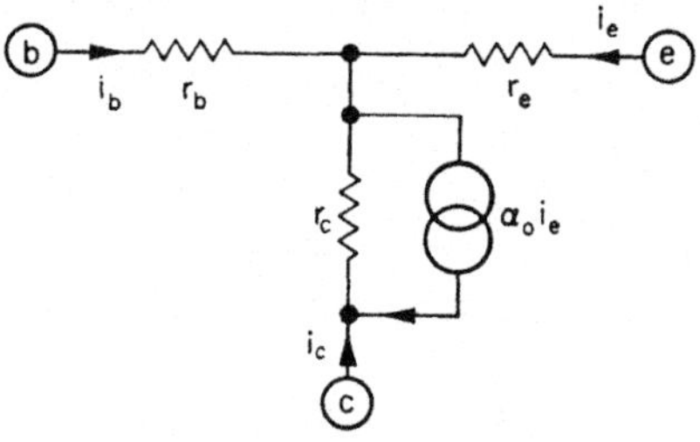

FIG. 4.17. Simplified grounded collector equivalent circuit.

4.6. *h* Parameters

For this and subsequent equivalent circuits the transistor is regarded as a " black box " with the input affecting the output and vice versa. Figure 4.18 shows the relationship for the common base configuration. The parameters h_{11}, h_{12}, h_{22} and h_{21} are defined in terms of the alternating components (v_1, i_1, v_2 and i_2)

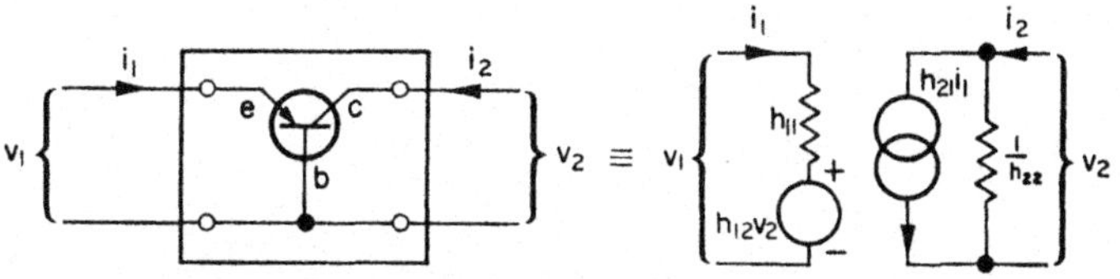

FIG. 4.18. Equivalent circuit using *h* parameters.

superimposed on the direct voltages and currents (V_1, I_1, V_2 and I_2), but they may also be stated in terms of small changes in the direct components themselves (δV_1, δI_1, δV_2, δI_2):

h_{11} = input impedance with collector short-circuited

$$= \left(\frac{v_1}{i_1}\right)_{v_2 = 0},$$

or slope of input characteristic

$$\left(\frac{\delta V_1}{\delta I_1}\right)_{V_2 = \text{constant}}$$

h_{12} = voltage feedback ratio with input open-circuited

$$= \left(\frac{v_1}{v_2}\right)_{i_1 = 0},$$

or slope of feedback characteristic

$$\left(\frac{\delta V_1}{\delta V_2}\right)_{I_1 = \text{constant}}.$$

h_{22} = output admittance with input open-circuited

$$= \left(\frac{i_2}{v_2}\right)_{i_1 = 0},$$

or slope of output characteristic

$$\left(\frac{\delta I_2}{\delta V_2}\right)_{I_1 = \text{constant}}.$$

h_{21} = forward current gain with collector short-circuited

$$= \left(\frac{i_2}{i_1}\right)_{v_2 = 0},$$

or slope of transfer characteristic

$$\left(\frac{\delta I_2}{\delta I_1}\right)_{V_2 = \text{constant}}.$$

These definitions mean that h parameters may be determined practically from the circuits shown in Fig. 4.19 or the static characteristics shown in Fig. 4.20. The static characteristics for a *pnp* transistor are shown with negative axes corresponding to negative voltages and current and not the more common method shown in the previous chapter, where only the moduli of the voltages and currents are considered.

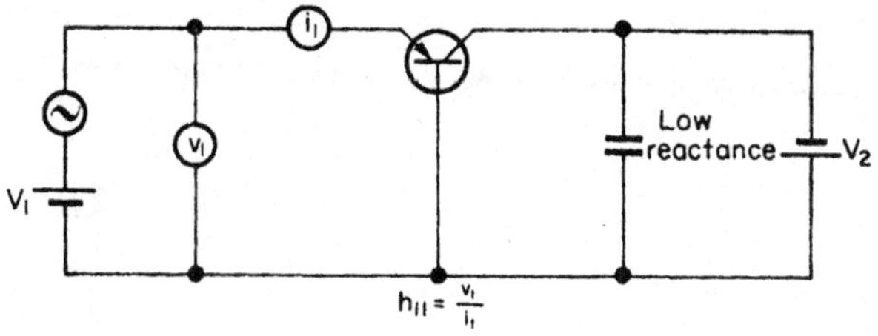

(a)

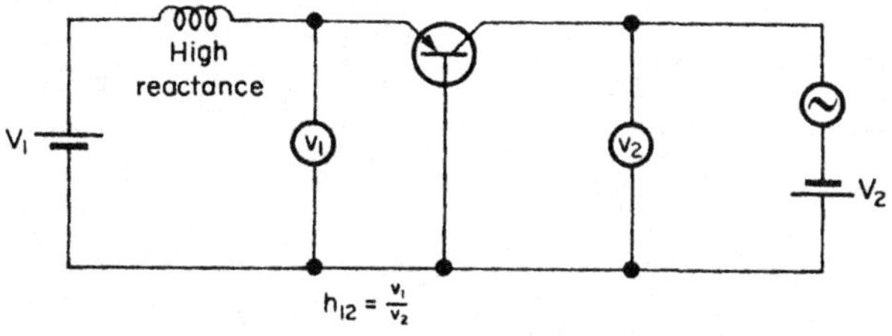

(b)

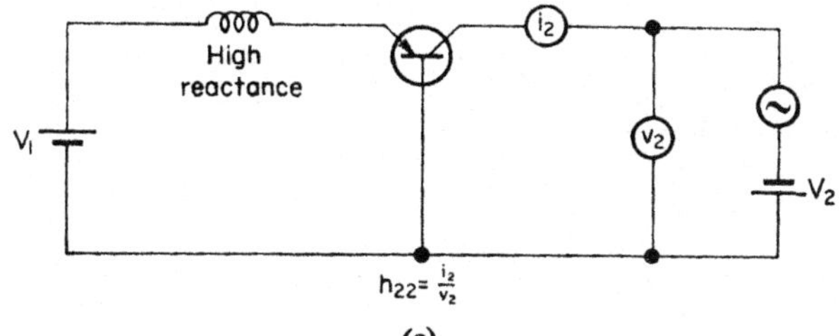

(c)

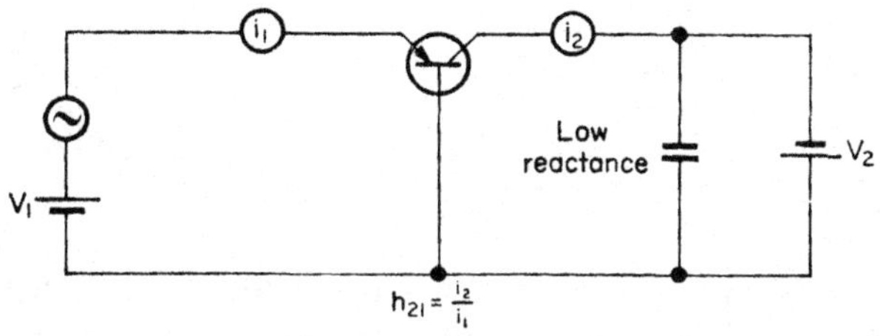

(d)

FIG. 4.19. Measurement of *h* parameters.

The values of h_{11}, h_{12} and h_{22} are positive but from Fig. 4.20(d), h_{21} is negative, but this is to be expected since currents flowing into a transistor are conventionally taken as positive, and for

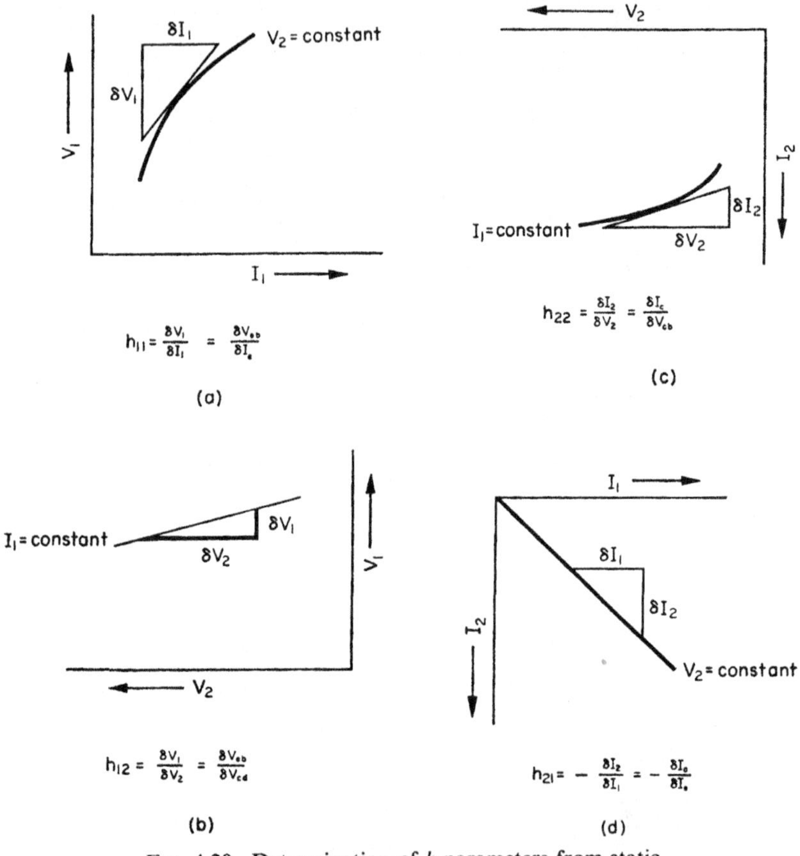

FIG. 4.20. Determination of h parameters from static characteristics.

this configuration an increase in current flowing into the transistor leads to an increase in current flowing out of the transistor. The static characteristics are not suitable for high frequencies where the h parameters may be complex.

The common emitter configuration leads to four similar parameters h'_{11}, h'_{12}, h'_{22} and h'_{21}, and the common collector gives the parameters h''_{11}, h''_{12}, h''_{22} and h''_{21} but sometimes an entirely different set of subscripts are used as shown in Table 4.1.

TABLE 4.1. ALTERNATIVE DESIGNATION OF h PARAMETERS

Input impedance	Output admittance	Reverse voltage gain	Forward current gain
$h_{11} = h_{ib}$	$h_{22} = h_{ob}$	$h_{12} = h_{rb}$	$h_{21} = h_{fb}$
$h'_{11} = h_{ie}$	$h'_{22} = h_{oe}$	$h'_{12} = h_{re}$	$h'_{21} = h_{fe}$
$h''_{11} = h_{ic}$	$h''_{22} = h_{oc}$	$h''_{12} = h_{rc}$	$h''_{21} = h_{fc}$

TABLE 4.2. COMPARISON OF h AND T PARAMETERS

Common emitter		Common collector	
h'_{11}	$r_b + r_e/1 - \alpha_0$	h''_{11}	r_b
h'_{12}	$r_e/r_c(1 - \alpha_0)$	h''_{12}	1
h'_{22}	$1/r_c(1 - \alpha_0)$	h''_{22}	$1/r_c(1 - \alpha_0)$
h'_{21}	$\dfrac{\alpha_0}{1 - \alpha_0} = \beta_0$	h''_{21}	$= \dfrac{1}{1 - \alpha_0} = -\gamma_0$

4.7. Relation between h and T Parameters

The simple low frequency T and h parameters may be compared by writing corresponding equations and equating coefficients for Figs. 4.21 and 4.22.

Figure 4.21 gives

$$v_1 = i_e(r_e + r_b) + i_c r_b$$

$$v_2 = i_c(r_c + r_b) + i_e(r_b + \alpha_0 r_c)$$

$$= i_c r_c + i_e \alpha_0 r_c \quad \text{(very nearly).}$$

Figure 4.22 gives

$$i_c = v_2 h_{22} + i_e h_{21}$$

or

$$v_2 = \frac{i_c}{h_{22}} - \frac{i_e h_{21}}{h_{22}},$$

$$v_1 = i_e h_{11} + h_{12} v_2$$

$$= i_e \left(h_{11} - \frac{h_{12} h_{21}}{h_{22}} \right) + i_c \frac{h_{12}}{h_{22}}.$$

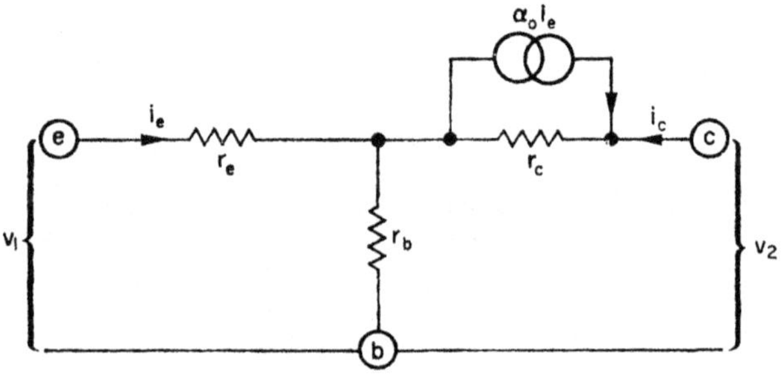

FIG. 4.21. Currents and voltages for a grounded base T equivalent circuit.

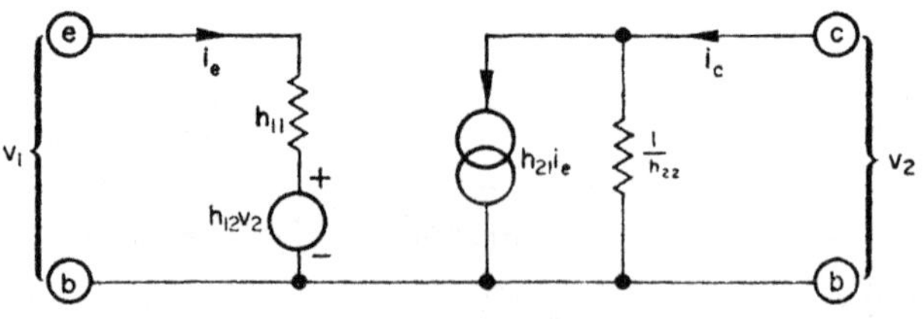

FIG. 4.22. Currents and voltages for grounded base h equivalent circuit.

Hence comparing coefficients gives

$h_{22} = 1/r_c$, $h_{21} = -h_{22}\alpha_0 r_c = -\alpha_0$, $h_{12} = r_b h_{22} = r_b/r_c$

$h_{11} = r_c + r_b + h_{12} h_{21}/h_{22} = r_e + r_b(1-\alpha_0)$.

The common emitter and common collector hybrid parameters may also be related to the T parameters, the results, after simplifying assumptions are made, being shown in Table 4.2.

4.8. Z Parameters

As in the case of h parameters, the Z parameters are determined by considering the transistor as a " black box ". Figure 4.23 shows the relationship for the common emitter configuration.

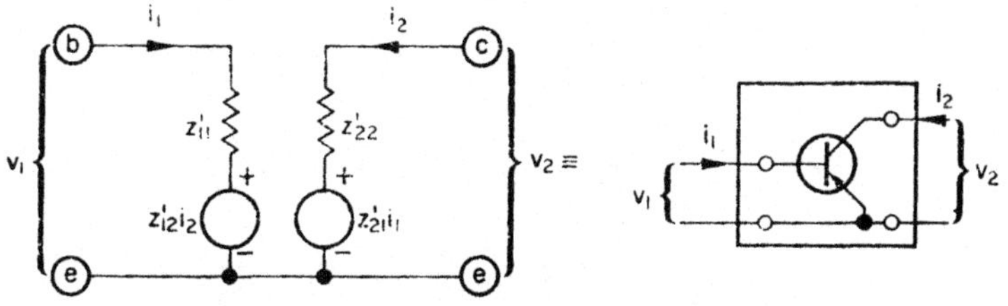

FIG. 4.23. Equivalent circuit using Z parameters.

The values of the Z parameters may be determined from circuits similar to those shown in Fig. 4.20:

$$Z'_{11} = \left(\frac{v_1}{i_1}\right)_{i_2=0} = \left(\frac{v_{be}}{i_b}\right) \quad \text{collector O.C.}$$

$$Z_{12} = \left(\frac{v_1}{i_2}\right)_{i_1=0} = \left(\frac{v_{be}}{i_c}\right) \quad \text{base O.C.}$$

$$Z_{22} = \left(\frac{v_2}{i_2}\right)_{i_1=0} = \left(\frac{v_{ce}}{i_c}\right) \quad \text{base O.C.}$$

$$Z_{21} = \left(\frac{v_2}{i_1}\right)_{i_2=0} = \left(\frac{v_{ce}}{i_b}\right) \quad \text{collector O.C.}$$

Similar equations enable the Z parameters to be evaluated for the common base and common collector configurations.

4.9. Relation between *Z* and *T* Parameters

Figures 4.23 and 4.24 enable comparison to be made between the *Z* and *T* parameters.

Figure 4.23 gives

$$v_1 = i_1 Z'_{11} + i_2 Z'_{12},$$

$$v_2 = i_2 Z'_{22} + i_1 Z'_{21}.$$

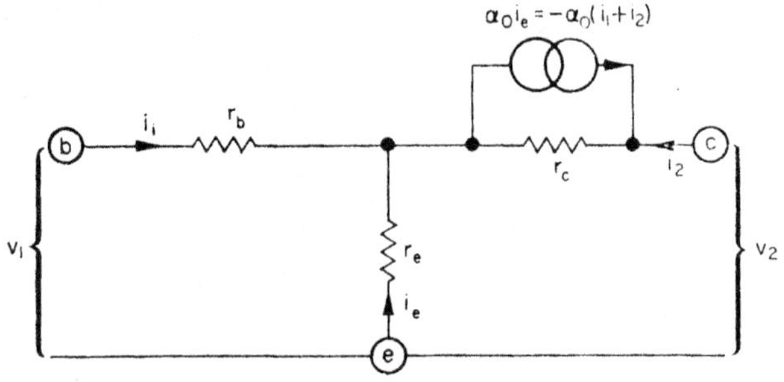

FIG. 4.24. Currents and voltages for a grounded emitter *T* equivalent circuit.

Figure 4.24 gives

$$v_1 = i_1(r_e + r_b) + i_2 r_e,$$

$$v_2 = i_2(r_e + r_c(1-\alpha_0)) + i_1(r_e - \alpha_0 r_c)$$

$$= i_2 r_c(1-\alpha_0) - i_1 \alpha_0 r_c \quad \text{(very nearly).}$$

Hence comparing coefficients gives

$$Z'_{12} = r_e, \quad Z'_{11} = r_e + r_b, \quad Z'_{21} = -\alpha_0 r_c, \quad Z'_{22} = r_c(1-\alpha_0).$$

The corresponding values for common base and common collector circuits are shown in Table 4.3.

4.10. *Y* Parameters

This set of parameters comprises four admittance parameters and is shown in Fig. 4.25 for the common collector configuration.

TABLE 4.3. COMPARISON OF Z AND T PARAMETERS

Common base		Common collector	
Z_{11}	$r_e + r_b$	Z''_{11}	r_c
Z_{12}	r_b	Z''_{12}	$r_c(1-\alpha_0)$
Z_{21}	$\alpha_0 r_c$	Z''_{21}	r_c
Z_{22}	r_c	Z''_{22}	$r_c(1-\alpha_0)$

The parameters are defined in a similar manner to those in the previous circuits:

$$Y''_{11} = \left(\frac{i_1}{v_1}\right)_{v_2=0} = \left(\frac{i_b}{v_{bc}}\right) \quad \text{emitter S.C.}$$

$$Y''_{12} = \left(\frac{i_1}{v_2}\right)_{v_1=0} = \left(\frac{i_b}{v_{ec}}\right) \quad \text{base S.C.}$$

$$Y''_{22} = \left(\frac{i_2}{v_2}\right)_{v_1=0} = \left(\frac{i_e}{v_{ec}}\right) \quad \text{base S.C.}$$

$$Y''_{21} = \left(\frac{i_2}{v_1}\right)_{v_2=0} = \left(\frac{i_e}{v_{bc}}\right) \quad \text{emitter S.C.}$$

Similar equations may be obtained for the common base and common emitter configurations.

4.11. Relation between Y and T Parameters

Figure 4.25 gives the circuit relations for the Y parameters and Fig. 4.26 for the T parameters.

Figure 4.25 gives

$$i_1 = v_1 y''_{11} + v_2 y''_{12},$$

$$i_2 = v_2 y''_{22} + v_1 y''_{21}.$$

Figure 4.26 gives

$$v_1 = i_1(r_b+r_c)+i_2r_c(1-\alpha_0),$$

$$v_2 = i_1r_c+i_2[r_e+r_c(1-\alpha_0)].$$

The latter equations may be rearranged to give

$$i_1 = \frac{v_1[r_e+r_c(1-\alpha_0)]}{r_b[r_e+r_c(1-\alpha_0)]+r_cr_e} - \frac{v_2r_c(1-\alpha_0)}{r_b[r_e+r_c(1-\alpha_0)]+r_cr_e},$$

$$i_2 = \frac{v_2(r_b+r_c)}{r_b[r_e+r_c(1-\alpha_0)]+r_er_b} - \frac{v_1r_c}{r_b[r_e+r_c(1-\alpha_0)]+r_er_b}.$$

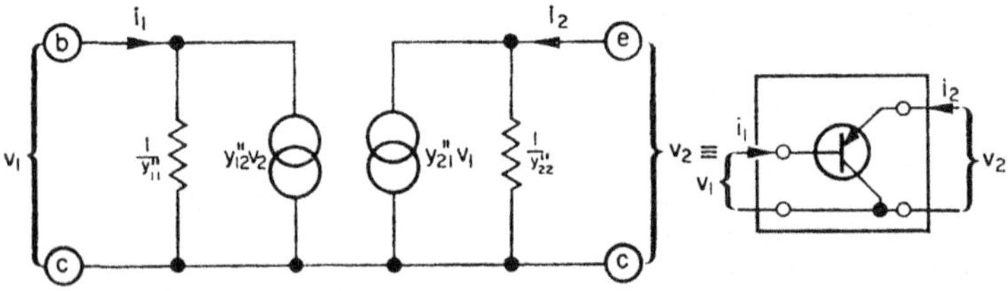

FIG. 4.25. Equivalent circuit using Y parameters.

Hence, since r_e is very much less than $r_c(1-\alpha_0)$ and r_b is very much less than r_c, comparing coefficients gives

$$Y''_{11} = \frac{1}{r_b+r_e/(1-\alpha_0)},$$

$$Y''_{12} = -\frac{1}{r_b+r_e/(1-\alpha_0)},$$

$$Y''_{22} = \frac{1}{r_b(1-\alpha_0)},$$

$$Y''_{21} = -\frac{1}{r_b(1-\alpha_0)}.$$

The corresponding values for common base and common emitter low frequency circuits are shown in Table 4.4, but the main application of this equivalent circuit is at high frequencies when the parameters become complex.

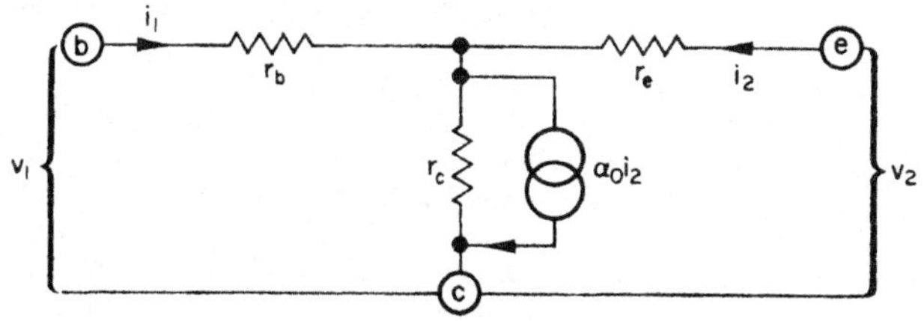

FIG. 4.26. Currents and voltages for a common collector *T* equivalent circuit.

TABLE 4.4. COMPARISON OF *Y* AND *T* PARAMETERS

	Common base		Common emitter
Y_{11}	$\dfrac{1}{r_e + r_b(1-\alpha_0)}$	Y'_{11}	$\dfrac{(1-\alpha_0)}{r_e + r_b(1-\alpha_0)}$
Y_{12}	$-\dfrac{r_b}{r_c[r_e + r_b(1-\alpha_0)]}$	Y'_{12}	$\dfrac{r_e}{r_c[r_e + r_b(1-\alpha_0)]}$
Y_{22}	$\dfrac{(r_e + r_b)}{r_c[r_e + r_b(1-\alpha_0)]}$	Y'_{22}	$\dfrac{r_e + r_b}{r_c[r_e + r_b(1-\alpha_0)]}$
Y_{21}	$\dfrac{\alpha_0}{r_e + r_b(1-\alpha_0)}$	Y'_{21}	$\dfrac{\alpha_0}{r_e + r_b(1-\alpha_0)}$

Questions for Chapter 4

Q.4.1. Prove the equations relating *h* and *T* parameters shown in Table 4.2.

Q.4.2. Prove the equations relating *Z* and *T* parameters shown in Table 4.3.

Q.4.3. Prove the equations relating *Y* and *T* parameters shown in Table 4.4.

Q.4.4. The parameters of a certain transistor are $r_e = 50\ \Omega$, $r_c = 1\ \text{M}\Omega$, $\alpha_0 = 0{\cdot}98$ and $r_b = 1\ \text{K}\Omega$. Determine the low frequency *h* parameters for common base, common emitter and common collector configurations.

Q.4.5. Using the same parameters as given in question 4.4, evaluate the low frequency Z parameters for the three possible transistor configurations.

Q.4.6. Using the same parameters as given in question 4.4, evaluate the low frequency Y parameters for the three possible transistor configurations.

Q.4.7. Draw an equivalent circuit, valid at low frequencies for a junction transistor used as a small signal amplifier, in the common emitter connection. Explain briefly the significance of the parameters used. One of two major factors is likely to limit the h.f. response in this connection, depending on the magnitude of the load resistance. State what these factors are, and give a simplified analysis of the effect of either one of them justifying any assumptions made. (I.E.E., June 1958)

PART II

Design of Circuits

CHAPTER 5

Rectifiers and Stabilizers

5.1. Semiconductor Diodes

A wide range of semiconductor diodes is available, made from a number of materials and by a variety of processes. The applications for semiconductor diodes cover the whole electromagnetic spectrum from ultraviolet frequencies to power rectification giving direct current.

The copper oxide diode, where the boundary between copper and cuprous oxide forms the rectifying junction, has been available for over thirty years. It is now only used as a low power device in rectifier instruments and as a modulator in multi-channel telephone systems.

The selenium rectifier has a junction consisting of selenium and a tin–cadmium alloy. Due to the low cost of manufacture, selenium rectifiers give reasonable efficiency for power outputs up to 10 kW and have been used in installations up to 1 MW. Both copper oxide and selenium cells can operate at maximum current densities of 100 mA/cm^2 of junction area and may be used in ambient temperatures of up to 55°C. Due to the relatively large mass of metal in these rectifiers, the copper or selenium device may be safely operated at ten times full load for some seconds. It is worth noting that if a selenium rectifier is operated with a steady current in the reverse direction beyond the breakdown voltage, then a lethal gas is produced.

Both point-contact and junction germanium diodes are available, and the junction types are usually made by alloying. The point-contact diode is very widely used as a low-cost device for currents of up to 30 mA at about 100 V. The runaway point for excessive leakage current occurs at peak inverse voltages in the range 500–700 V (d.c.) for germanium junction rectifiers and the maximum permissible junction temperature is about 90°C.

The silicon crystal diode has been available for many years as a mixer and detector in microwave radar and communication systems. The subminiature silicon *pn* junction gives superior performance to the point-contact germanium diode, and may ultimately replace that device in many lower power applications.

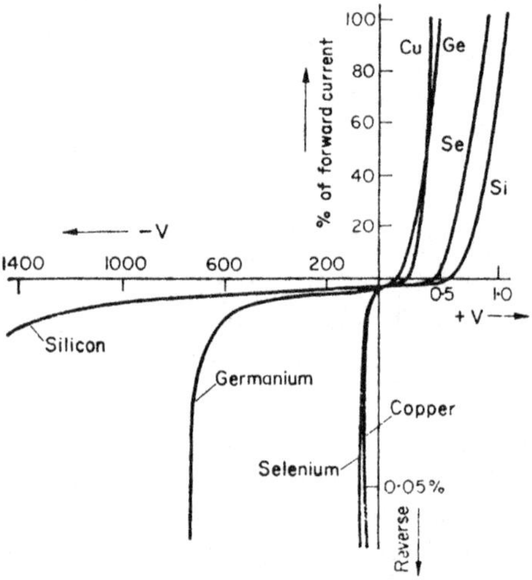

FIG. 5.1. Forward and reverse characteristics of various semiconductor rectifiers.

Silicon junction rectifiers may be made equally well by either diffusion or alloying. The current/voltage characteristics of these rectifiers are excellent and it is unlikely that any new combination of materials will offer any significant improvement. Both silicon and germanium can carry currents up to the order of 300 A/cm^2 of junction area. This is about the same maximum current density as in a copper conductor. The peak inverse voltage may be made of the order of 2 kV for a silicon junction diode and, if high temperature solders are used in the construction, a junction temperature of 200°C may be approached. When calculating the power dissipation in a germanium diode the reverse leakage current must be taken into account, but for the silicon diode only the product of the forward drop and the forward current need be considered. The graphs of Fig. 5.1 show the forward and reverse characteristics of various rectifiers, based on d.c. measurements.

Figure 5.1 shows clearly that the germanium junction diode starts to conduct in the forward direction, at a lower voltage than the equivalent silicon diode. This means that at direct voltages below 100 V and with large currents the germanium device is

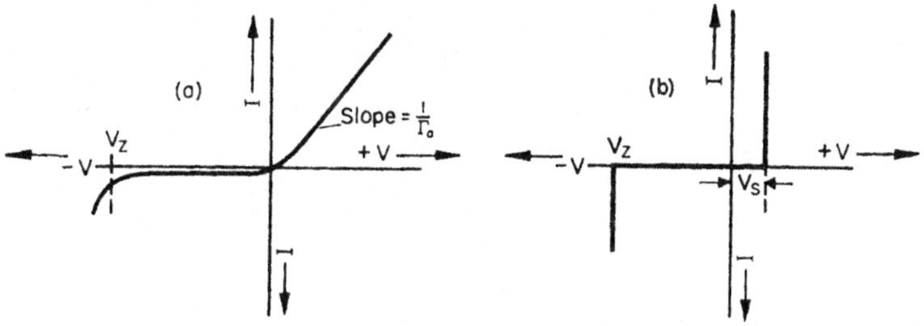

FIG. 5.2. Simplified diode characteristics.

more efficient than silicon. The simplified characteristics of Fig. 5.2 may be useful for calculations. Graph (a) shows a device, such as a germanium point-contact diode, which starts to conduct at approximately zero voltage and thereafter has a forward slope resistance of r_a ohms. On the other hand, graph (b) shows a device, such as a silicon junction diode, which conducts at a forward starting voltage of V_s volts, and beyond that voltage has a negligible forward slope resistance. For a typical silicon diode the starting voltage V_s is about 700 mV.

The silicon reference (or " zener ") diode is manufactured from low resistivity material to give a sharply defined reverse breakdown voltage. A wide range of breakdown voltages from about -2 V, and permissible inverse dissipations, to about 50 W, are available.

The silicon controlled rectifier (or " thyristor "), which is a four-layer *pnpn* device, with a third, or " gate ", contact is rapidly coming into use in applications where control for power variation or regeneration is required. The S.C.R. is considered in more detail in Sections 5.9 and 5.10.

An application of the tunnel diode is considered in Chapter 8, the photo-electric *pn* diode is discussed in Chapter 10 and the small *pnpn* diode, without gate control, is mentioned in Chapter 11.

5.2. Half-wave Rectifier

The half-wave circuit, illustrated in Fig. 5.3, is simple, in that there is only one diode, and an intermediate transformer is not required in some applications. However, the use of this circuit is restricted to lower powers, as the overall efficiency is low and the

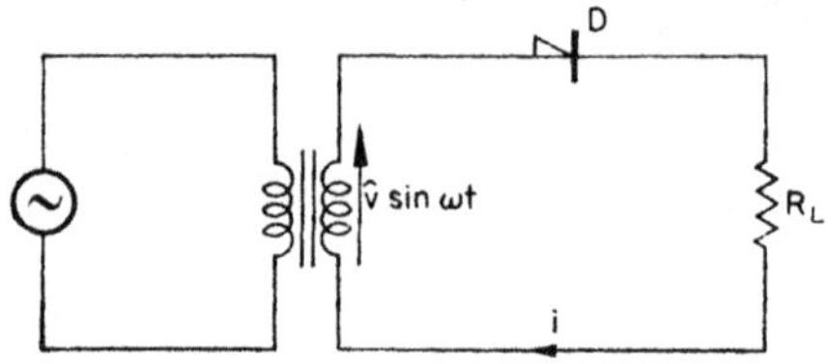

FIG. 5.3. Half-wave rectifier circuit.

output waveform has a large proportion of ripple. The circuit may be analysed in the following manner.

If the input voltage across the transformer secondary is $\hat{v}\sin\omega t$ and the diode has a forward slope resistance of r_a ohms then the instantaneous current is,

$$i = \frac{\hat{v}\sin\omega t}{r_a + R_L}, \quad \text{in the positive half-cycle,}$$

and

$i = 0$, in the negative half-cycle, with respect to the input waveform.

The mean current over a whole cycle of the input wave is given by

$$i_m = \int_0^{\pi} \frac{\hat{v}\sin\omega t}{(r_a+R_L)}\frac{d(\omega t)}{2\pi} = \frac{\hat{v}}{\pi(r_a+R_L)}.$$

At the same time the root mean square (r.m.s.) current is,

$$i_{r.m.s.} = \left[\frac{1}{2\pi}\int_0^{2\pi}\left(\frac{\hat{v}\sin\omega t}{r_a+R_L}\right)^2 d(\omega t)\right]^{1/2} = \frac{\hat{v}}{2(r_a+R_L)}.$$

Now the input power from the transformer is given by

$$P_{IN} = (i_{r.m.s.})^2\,(r_a+R_L).$$

The maximum available direct output power is

$$P_{OUT} = (i_m)^2 R_L.$$

The difference between the input and output powers represents the dissipation in the diode and the ripple power in the output. With ideal smoothing and no ripple in the output, the modified input power would be given by

$$P'_{IN} = (i_{r.m.s.})^2 r_a + (i_m)^2 R_L.$$

Hence the efficiency of a.c. input to d.c. output power is

$$\eta = \frac{4}{\pi^2}\left(\frac{R_L}{r_a+R_L}\right) 100\%.$$

If we assume that the diode has a forward starting voltage V_s, and zero forward resistance, then this will have the effect of giving an angle of flow of less than 180° in the conduction half-cycle. So the instantaneous current flowing is then

$$i = \frac{\hat{v}\sin\omega t - V_s}{R_L}$$

and the power loss in the diode is equivalent to $i_{r.m.s.}V_s$. This type of diode will be a more efficient rectifier as soon as V_s is less than 1% of $\hat{v}$.

5.3. Full-wave Rectifier

The full-wave rectifier circuit with a centre-tapped transformer and two thermionic diodes is in extensive use. This consists essentially of two half-wave circuits rectifying in push-pull on

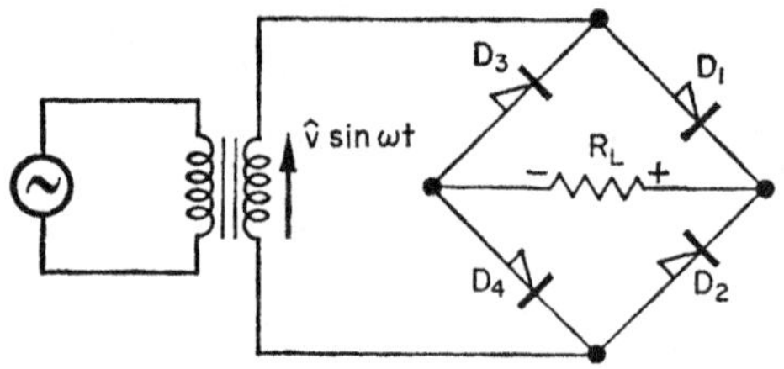

FIG. 5.4. Bridge rectifier circuit with four diodes.

alternate half-cycles of the input waveform. Since the single-phase supply may then be considered as split into two phases, this is sometimes known as a biphase rectifier. However, with semiconductor diodes the centre-tapped transformer circuit is not widely used.

The single-phase rectifier using a diode bridge, see Fig. 5.4, shows several distinct advantages over the centre-tap circuit. Since two diodes are at all times in series, the d.c. output voltage, for a given diode, is twice that of the centre-tap circuit. The transformer is more simple to manufacture and smaller, as current is flowing continuously through the complete secondary winding over the whole cycle of voltage. In fact, each diode symbol of Fig. 5.4 may represent a series–parallel combination of diode units.

When semiconductor diodes are joined in series to obtain higher reverse voltage strength, it is essential to connect voltage-sharing resistors in parallel. The characteristics of individual diodes will differ in both reverse resistance and carrier storage time. Without shunting resistors the diode with the highest reverse resistance might be broken down by excessive peak inverse voltage drop. In the example shown in Fig. 5.5 the four diodes are a low voltage germanium type, with a breakdown voltage of −45 V and a reverse resistance in the range 1 to 5 MΩ. To avoid much ripple and yet remain effective, the voltage-sharing resistor must be approximately the geometric mean value of the forward

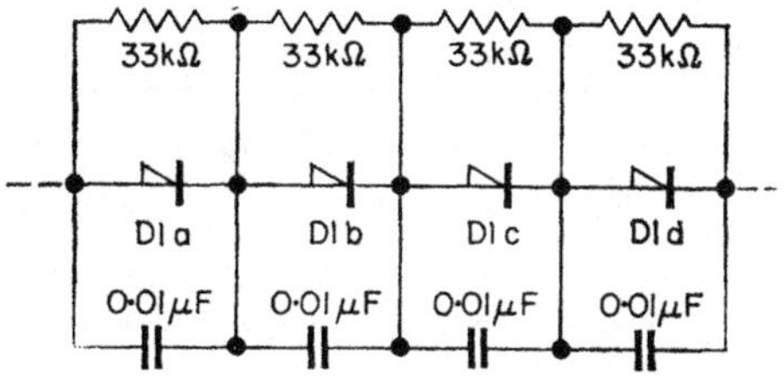

FIG. 5.5. Example of voltage sharing with diodes in series.

and reverse resistances of the diode. To overcome the differences in carrier storage time it may also be necessary to add parallel capacitors in addition to the voltage-sharing resistors. An added shunt capacitance of 0·1 μf per diode is commonly used for currents of 100 mA to 1 A.

The power efficiency of the bridge rectifier depends on the resistance, or volt drop, of at least two diodes in series. Thus the instantaneous forward current for a four-diode bridge will be given by

$$\text{(a)}\ i = \frac{\hat{v}\sin\omega t}{R_L + 2r_a} \quad \text{or} \quad \text{(b)}\ i = \frac{\hat{v}\sin\omega t - 2V_s}{R_L},$$

where r_A is the forward resistance and V_s the forward volt drop across each rectifier.

A single-phase rectifier, as used for low or medium power applications, usually incorporates a smoothing filter circuit, consisting of one or more reactances. A single shunt capacitor is satisfactory for low current applications, but has poor regulation for a varying current load. The choke input filter presents a series inductance to the rectifier, and is often a disadvantage due to its weight and bulk. The stabilizer circuits discussed later in this chapter also help to remove voltage ripple and the design of reactive filters will not be considered in detail.

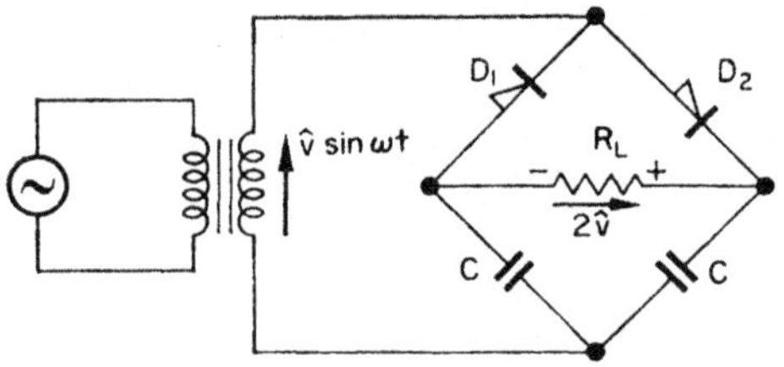

FIG. 5.6. Voltage doubler circuit.

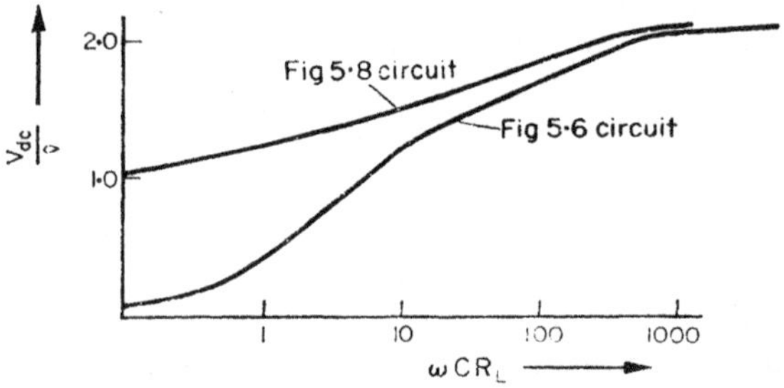

FIG. 5.7. Regulation curve for simple voltage doubler.

5.4. Voltage Multipliers

Voltage multiplying rectifier circuits find application as a source of direct voltage at very small currents, since no standby power is consumed. This is the case in telemetry systems, where the remote apparatus may be controlled by rectification of audio frequency tones, and with some cathode ray tubes as the E.H.T. source.

A voltage multiplier uses the diode as a switch to charge capacitors in parallel and obtain a higher direct voltage from these capacitors in series. The most common voltage doubler is shown in Fig. 5.6. If the input waveform has a peak voltage of $\hat{v}$ and no current is drawn from the circuit, then one capacitor charges to $+\hat{v}$ and the other capacitor to $-\hat{v}$. Since this circuit is similar in operation to the full-wave rectifier, the ripple voltage is at twice the supply frequency. The capacitors must be rated at

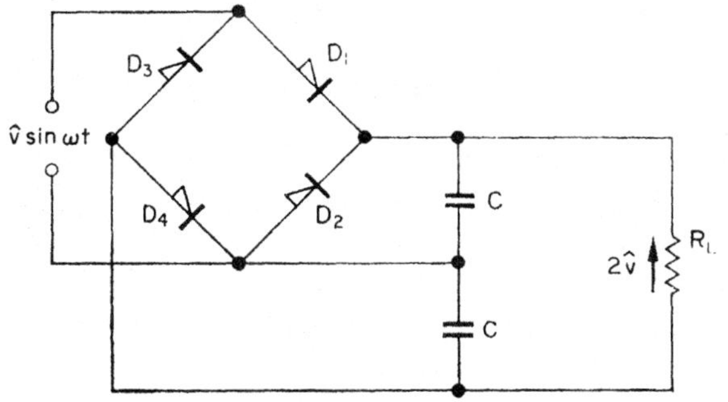

FIG. 5.8. Bridge voltage doubler circuit.

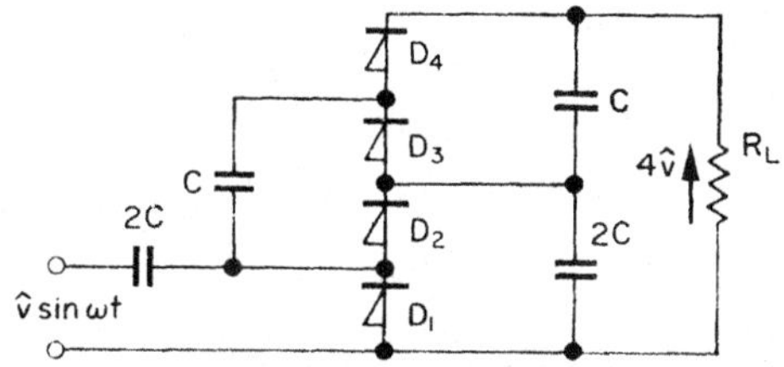

FIG. 5.9. Voltage quadrupler circuit.

the a.c. peak voltage $\hat{v}$ whilst the maximum inverse voltage across each diode is $2\hat{v}$. The regulation of the circuit is poor, as indicated by the graph of Fig. 5.7. A time constant of $\omega CR > 400$ is desirable to give low ripple and approximate voltage doubling.

The bridge voltage doubler of Fig. 5.8 gives a better regulation of output voltage for changes of load current, since it acts simply as a bridge rectifier for higher load currents, Additionally, this circuit imposes a lower peak inverse voltage on each diode.

A general form of voltage multiplier, with one common input and output terminal, can be based on the quadrupler of Fig. 5.9. In this circuit each capacitor and diode must be rated at twice the peak input voltage. The practical limit is reached at a voltage multiplication of ten times, for which the input capacitance must be large and the voltage regulation will be very poor.

5.5. Diode Voltmeters

A diode rectifier can be used with a microammeter, in the circuit of Fig. 5.10, as an a.c. voltmeter and this principle is widely applied. In order that the capacitor C has very little time to discharge between current pulses from the diode D, the time

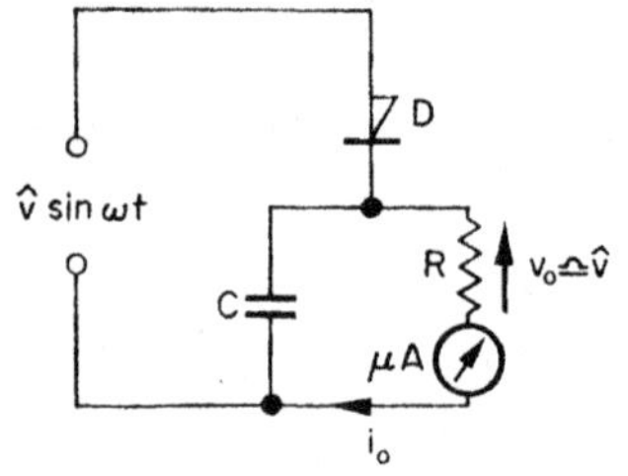

FIG. 5.10. Simple peak-reading diode voltmeter.

constant of the circuit should be large. If the periodic time of the input wave is T sec, then we must have $CR > 100T$. With this condition fulfilled the potential across the capacitor is almost the peak input voltage, and the microammeter reads a steady current i_0, where

$$\hat{v} = i_0 R \text{ volts.}$$

In many commercial instruments the diode detector stage is followed by d.c. amplification, since the microammeter is a delicate instrument. The diode and the associated R.C. circuit may be in a probe unit, joined to the main voltmeter by a screened cable. The d.c. amplifier is commonly a bridge circuit with a milliammeter in the detector arm. Since this milliammeter is calibrated in r.m.s. values, and the diode detector gives a voltage proportional to peak input, so the meter reading is correct only when the input voltage has a sinusoidal waveform.

A voltmeter calibrated to read "true" r.m.s. voltages is accurate for any waveform, however random (such as electrical noise), by incorporating a circuit having a true square law characteristic. A

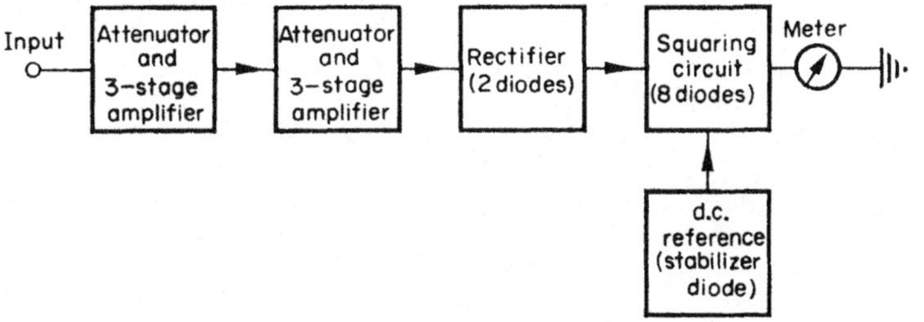

FIG. 5.11. Schematic for a voltmeter reading true r.m.s. values.

typical true r.m.s. voltmeter is illustrated by the block schematic of Fig. 5.11. This consists of two stages, each of a high impedance attenuator followed by a wide-band amplifier, with a generous amount of negative feedback. The output from the final amplifier is transformer coupled to a full-wave rectifier, which feeds a squaring circuit. This squaring circuit consists of eight germanium point-contact diodes arranged, with suitable d.c. bias and different load resistors, to generate a square law function by a continuous sequence of linear segments. Logarithmic calibration of the output meter then provides an accuracy of about 1% and a linear decibel scale.

5.6. Diode Stabilizer

The reverse current/voltage characteristic of a silicon junction reference diode approximates to that of a voltage operated switch and is illustrated in Fig. 5.12. This class of diode, with a sharply defined voltage breakdown point, is commonly known as the zener diode, although the zener, or field, effect is really only relevant at voltages less than 5·8 V. Certainly, at reverse voltages greater than 8 V, the breakdown to easy conduction is due to avalanche, or electron multiplication, effect, and the higher voltage reference diodes are sometimes known as avalanche diodes. Provided that the available voltage is greater than that needed for reverse conduction, then the diode will conduct over a wide range of reverse current, with little voltage change (that is, the reverse

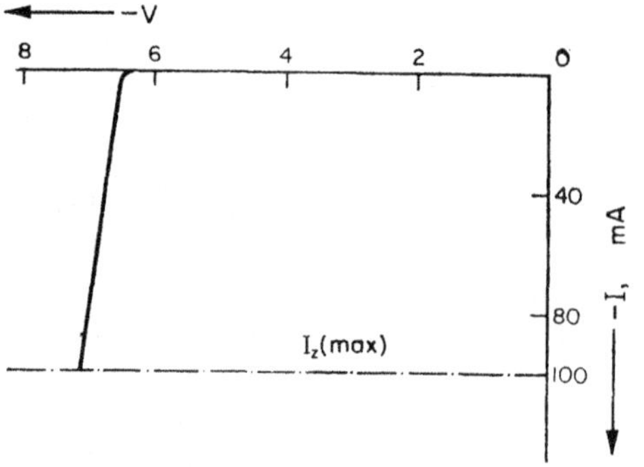

FIG. 5.12. Characteristic of small zener diode.

dynamic resistance is small). It is also possible to make a double-anode junction diode, which is essentially two zener diodes joined back to back. This device has a sharp breakdown to conduct large current at specified voltages both positive and negative, and is generally known as a regulator diode. This regulator may be formed by the diffusion of a *p* type impurity into an *n* type slice from both sides. Although this is a three-layer structure like a transistor, there is no contact to the centre layer and this middle region may be relatively wide.

The silicon zener diode may be used in voltage and current stabilizer circuits, having many advantages over gas-filled diodes.

The zener diode does not require a striking voltage greater than the running voltage and, in addition, is available in a very wide range of voltages and powers. For zener diodes with breakdown voltages (V_z) greater than 6 V, the zener voltage increases in magnitude with increase of junction temperature and this coefficient also increases with higher values of V_z. A small negative temperature coefficient exists for diodes with V_z less than 5 V, and this does lead to the possibility of obtaining an effectively zero temperature coefficient, with diodes in series, for the lower voltage applications. In addition, the actual value of reverse breakdown voltage differs from diode to diode of the same type

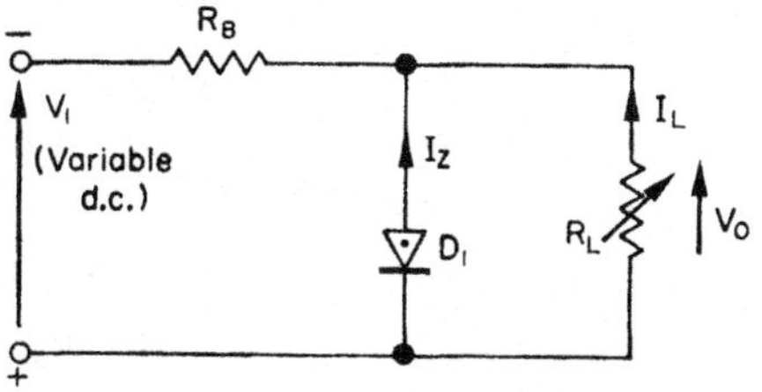

FIG. 5.13. Shunt diode stabilizer circuit.

and the diodes are usually available with ±5% or ±10% tolerance of V_z. The zener diode is also useful as a surge limiter and as a low resistance coupling device between points of different d.c. potential (as in d.c. amplifiers), but in this chapter we shall concentrate on the stabilizer circuit applications.

The basic shunt stabilizer circuit is shown in Fig. 5.13. To analyse this circuit we may assume that the zener diode has a reverse breakdown voltage of V_z volts and a slope resistance, when conducting, of R_z ohms, both of which are constant. The maximum permissible power dissipation of the diode limits the reverse current to I_{max} milliamps. The ballast resistor (R_B ohms) would be calculated knowing the input and output voltage conditions, the load current and the diode current. Using the symbols shown in Fig. 5.13, then the current flowing through the zener diode is $I_z = (1/R_z)(V_o - V_z)$ and the total supply current is $I_T = I_Z + I_L$. The input voltage is $V_{IN} = I_T R_B + V_O$.

Hence,

$$V_{IN} = (I_L + I_Z)R_B + V_O = \frac{(V_O - V_Z)}{R_Z} R_B + V_O + I_L R_B$$

or

$$V_{IN} = V_O\left(1 + \frac{R_B}{R_Z}\right) - V_Z\left(\frac{R_B}{R_Z}\right) + I_L R_B. \quad (1)$$

For an approximate result we may assume that V_Z and I_L are constant, and then differentiated, so that the differentials of V_Z and I_L are zero, to give,

$$\frac{\partial V_O}{\partial V_{IN}} = \frac{R_Z}{R_B + R_Z}. \quad (2)$$

Or, by rearranging eqn. (1), we have

$$I_L R_B = V_Z(R_B/R_Z) + V_{IN} - V_O(1 + R_B/R_Z). \quad (3)$$

Now assuming that V_{IN} is held constant and V_Z again remains constant, the relationship between a change in load current and output voltage follows from,

$$\frac{\partial V_O}{\partial I_L} = -\frac{R_B R_Z}{R_B + R_Z} \text{ ohms.} \quad (4)$$

Equation (2) shows how the output voltage varies with changes in input, assuming constant load. This ratio is often given the symbol S, where S = Voltage stability factor = $\partial V_O/\partial V_{IN}$.

Alternatively, the stabilization may be defined as the ratio of the percentage change in input voltage to the percentage change in output voltage.

Equation (4) gives the output resistance of the circuit, for a zero impedance source and constant input voltage. In general terms,

$$Z_O = \text{Output impedance} = \partial V_O/\partial I_L \text{ ohms.}$$

Also we must consider the effect of temperature changes, on the zener diode, and define,

$$K_T = \text{Temperature stability factor} = \partial V_O/\partial T \text{ volts.}$$

Then the total change in output voltage is given by

$$\partial V_O = S\partial V_{IN} + Z_O\partial I_L + K_T\partial T \text{ volts.}$$

For many calculations the shunt diode stabilizer may be replaced by the equivalent circuit of Fig. 5.14. In this circuit the battery has a voltage equal to the nominal breakdown voltage V_Z and the resistance in series with this battery is the approximate diode slope resistance R_Z. In fact a few stabilizer circuits, designed to operate with transistor equipment, did use mercury cells as the stabilizing element before zener diodes were widely

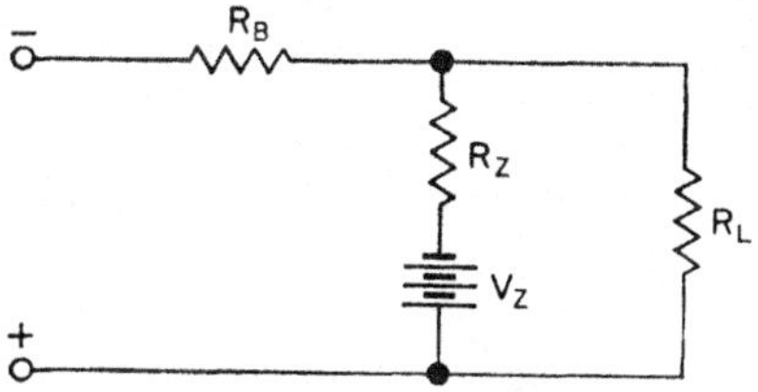

FIG. 5.14. Approximate equivalent of circuit in Fig. 5.13.

available. Then current only flows in the resistance R_Z when the output voltage to the shunt arm exceeds the "backing-off" voltage.

5.7. Transistor–Diode Stabilizers

The performance of power stabilizer circuits is greatly improved by the addition of amplifying devices, such as transistors, to the reference source. These circuits sometimes become very complex, particularly when the design takes account of wide temperature variations, but a simple stabilizer circuit, employing one series transistor, is shown in Fig. 5.15. The series transistor may be considered as a current divider so that only a small proportion of the load current, flowing into the emitter of the transistor, flows through the reference diode. This proportion is given by

$$\bar{\beta} = \frac{\Delta I_c}{\Delta I_B} \simeq \frac{\Delta I_E}{\Delta I_B}$$

At the same time, once the zener diode is conducting, this holds the base voltage of the transistor almost constant. If ideal semiconductor devices were available, the output voltage would be

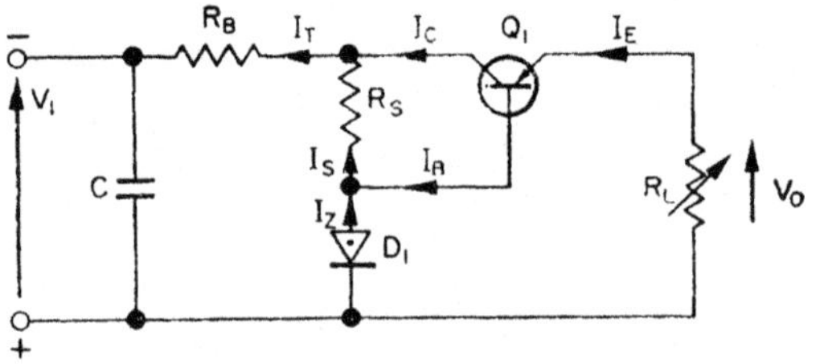

FIG. 5.15. Series transistor stabilizer.

constant regardless of changes in input voltage or of load current. In fact the base to emitter voltage V_{BE} and the voltage across the diode are dependent on the current flowing. As stated previously, the zener voltage (V_Z) and V_{BE} of the transistor are also temperature dependent.

The improvement of stability, output resistance and power efficiency associated with the circuit of Fig. 5.15 will only be obtained if certain precautions are taken. The maximum swing of input voltage for a given load current, must not exceed the voltage rating of the transistor. On the other hand, the actual output voltage may be much larger than the transistor voltage rating. There is considerable power dissipation in the transistors, and this often leads to the need for two or more transistors in parallel. Unless the transistor is to be destroyed by a short-circuit across the output terminals, a fast operating protection arrangement must be included.

A very approximate analysis, useful for comparison purposes, of the series transistor stabilizer may be made if we consider the transistor as a constant current device, with a mutual conductance of g_m amps/volt, a shunt resistance of r_c ohms and zero base current. The reference diode may be approximately considered as an ideal switch with zero slope resistance ($R_z = 0$). Then a change in load current can be calculated from

$$\partial I_L = g_m \partial V_{BE} + \frac{\partial V_{CE}}{r_c}$$

using the quantities indicated in Fig. 5.16. Either the load current or the input voltage may be taken as constant, to give an estimate of the voltage stability factor or the output resistance.

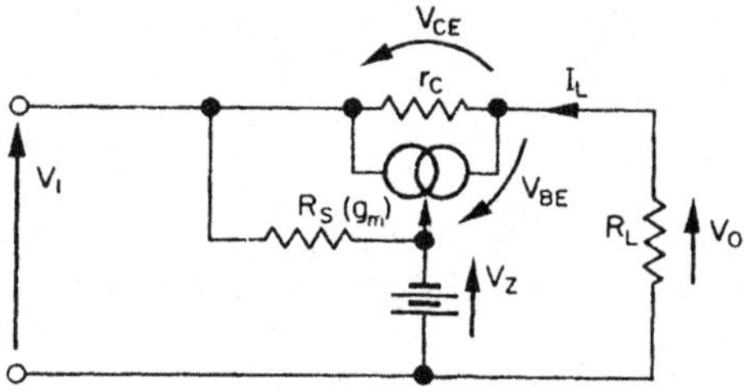

FIG. 5.16. Approximate equivalent of circuit in Fig. 5.15.

An alternative stabilizer circuit has the power transistor in shunt with the load, as illustrated in Fig. 5.17. The zener diode serves to hold the base–collector voltage of the transistor constant and this, in turn, stabilizes the load conditions. The shunt circuit suffers from the disadvantages that the main regulating current passes through the transistor, giving poor power efficiency, and that the collector–emitter voltage rating of the transistor must at least equal the output voltage. On the other hand, the transistor is not destroyed by an accidental short-circuit across the output terminals.

The simple stabilizer, such as that of Fig. 5.15, has an output resistance which is too large for some applications and an output voltage which is sensitive to temperature changes. The output impedance can be considerably reduced by using a feedback amplifier to control the series transistor. This feedback is provided by transistor Q_2 in Fig. 5.18. It is usually necessary to

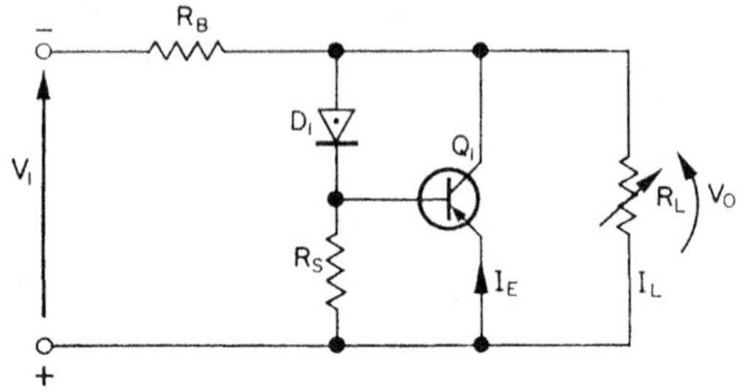

FIG. 5.17. Shunt transistor stabilizer.

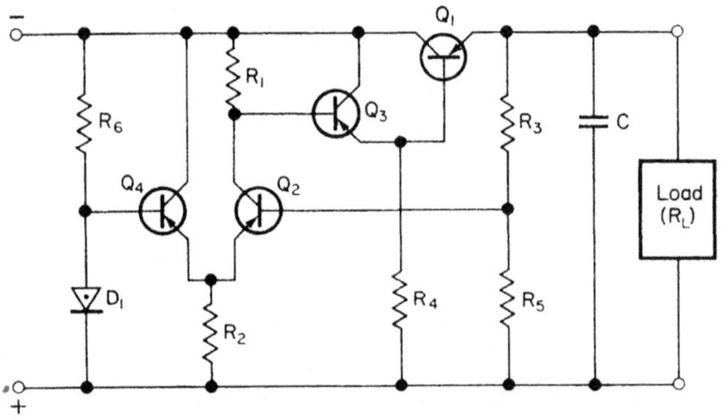

FIG. 5.18. Stabilizer with feedback and temperature compensation.

include another transistor (Q_3) in the feedback loop to reduce the current drawn from the collector of Q_2. Resistor R_4 is necessary to ensure that at very low output currents there is no chance of Q_3 being cut off by a reversal of the base current of Q_1. The temperature effects can be reduced by the use of two emitter-coupled transistors (a " long-tailed pair ") and this is incorporated in the circuit of Fig. 5.18. Here the reference voltage and a proportion of the output voltage are applied to the bases of Q_4 and Q_2. The voltage across R_2 is equal to $V_2 - V_{BE_4}$. Then, as the temperature increases V_{BE_4} will decrease by about 3 mV/°C and therefore the voltage across R_2 will increase by 3 mV/°C. This will provide the necessary reduction in the base–emitter voltage of Q_2 to ensure that the current in R_1 remains the same and the output voltage constant.

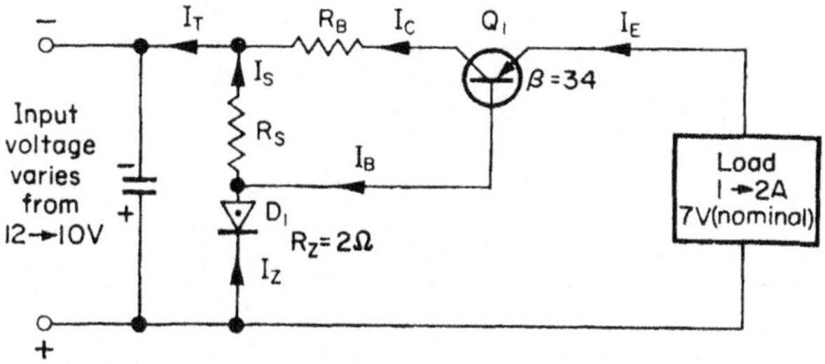

FIG. 5.19. Stabilizer circuit for design example.

5.8. Stabilizer Design Example

Since many of the properties of transistors and diodes have an effect on the successful operation of power stabilizers we shall consider a simple numerical design problem. This will show how the characteristics of the semiconductor device may be related to the practical performance of the circuit. The circuit for this example is that of Fig. 5.19, where the large input capacitance prevents the possibility of spurious oscillations.

The output characteristics for the power transistor Q_1 show that this has a current gain of $\beta = 34$, over the current range of interest. The base to emitter voltages, from the input characteristics, are found as

$$V_{BE} = 600 \text{ mV for } I_B = 60 \text{ mA}$$

and

$$V_{BE} = 500 \text{ mV for } I_B = 30 \text{ mA}.$$

The zener diode can conduct to a maximum of 100 mA. The zener voltage is $V_Z = 7{\cdot}6$ V at 5 mA and the diode slope resistance is approximately constant at $R_z = 2\ \Omega$ from 5 to 100 mA.

We wish to calculate suitable values for the resistors R_B and R_S, to estimate the stability and output resistance, and to find the maximum power dissipation at the various components. These calculations can be performed at the extreme conditions of input voltage and output current.

(a) 10 V in: 2 A out. We may assume that $I_E \simeq I_C$ and that

$$I_B = \frac{I_E}{\beta} = \frac{2000}{34} = 60 \text{ mA}.$$

Choosing I_Z at the minimum value of 5 mA (to avoid the curvature of the zener diode characteristic) gives $V_Z = 7{\cdot}6$ V. Hence the voltage drop across R_S is $V_S = 10 - 7{\cdot}6 = 2{\cdot}4$ V, where $I_S = I_B + I_Z = 65$ mA.

This suggests that a suitable value is $R_S = 2400/65 = 37\ \Omega$. Taking the nearest standard value, $R_S = 33\ \Omega$.

The base to emitter volt drop across the transistor for $I_E = 2$ A is $V_{BE} = 600$ mV. Therefore the output voltage is given by

$$V_O = V_Z - V_{BE} = 7{\cdot}6 - 0{\cdot}6 = 7{\cdot}0 \text{ V}.$$

If we allow a minimum of 1 V drop across the transistor, then the available p.d. for R_B is 2 V with 2 A flowing. So $R_B = 1\ \Omega$, with a power dissipation of 4 W.

(b) 12 V in: 1 A out. Now $I_B = 30$ mA (assuming β is unchanged). To this must be added the current of 5 mA, corresponding to $V_Z = 7{\cdot}6$ V. So the change in I_Z is given by

$$\Delta I_Z = \frac{12 - 7{\cdot}6 - 10^{-3}(33 \times 35)}{R_S + R_Z} = \frac{12 - 8{\cdot}76}{35} = 92 \text{ mA}.$$

Under this condition the voltage across the zener diode rises by 180 mV, whilst the value of $V_{BE} = 500$ mV. This results in an output voltage of $V_O = 7{\cdot}78 - 0{\cdot}5 = 7{\cdot}28$ V.

(c) 12 V in: 2 A out. Again $I_B = 60$ mA and this current, together with the minimum diode current of 5 mA, produces a p.d. of (65 × 33) mV across R_S. Hence

$$\Delta I_Z = \frac{12-7{\cdot}6-10^{-3}(65\times 33)}{35}\ \text{A} = \frac{2{\cdot}25}{35}\ 10^3\ \text{mA} = 64\ \text{mA}.$$

This results in $\Delta V_Z = 64\times 2\ \text{mV} \simeq 130$ mV, whilst $V_{BE} = 600$ mV. The output voltage is now given by

$$V_O = 7{\cdot}73-0{\cdot}6 = 7{\cdot}13\ \text{V}.$$

At the same time the voltage across the transistor is given by

$$12-2-7{\cdot}13 \simeq 2{\cdot}9\ \text{V} \quad (\text{say } 3\ \text{V}).$$

The maximum dissipation in the transistor is then slightly less than 6 W. The current through R_S, produced in condition (b), is

$$I_S = (92+5+30)\ \text{mA} = 127\ \text{mA}$$

and in condition (c), we have

$$I_S = 65+64 = 129\ \text{mA} \quad (\text{say } 130\ \text{mA}).$$

This last current gives a power dissipation of

$$P_{s(\max)} = 33(0{\cdot}13)^2 = 0{\cdot}56\ \text{W}.$$

Also, in condition (b) the maximum zener diode current is

$$I_S = 92+5 = 97\ \text{mA} \quad (\text{say } 100\ \text{mA}),$$

which corresponds to a volt drop of $V_Z = 7{\cdot}78$ V across the diode. Hence the maximum diode dissipation is 0·778 W, which leads to the requirement for a diode of 1 W rating.

For a load current increase of 1 A (with V_{IN} at 12 V), the output voltage changes from 7·28 to 7·13 V.

Therefore

$$R_{OUT} = \frac{7{\cdot}28-7{\cdot}13}{1-2} = -0{\cdot}15\ \Omega,$$

where the significance of the negative sign is that the output voltage falls for an increase in output current.

Also for a constant load current of 2 A and a change of input voltage from 10 to 12 V, the output voltage increases from 7·0 to 7·13 V. This corresponds to

$$S = \frac{0{\cdot}13}{2} = 0{\cdot}065 \quad \text{or} \quad 6{\cdot}5\%.$$

It is worth noting that an increase of current gain, which may be achieved by putting two transistors in series, would improve both stability and output resistance, and reduce the changes of current through the zener diode.

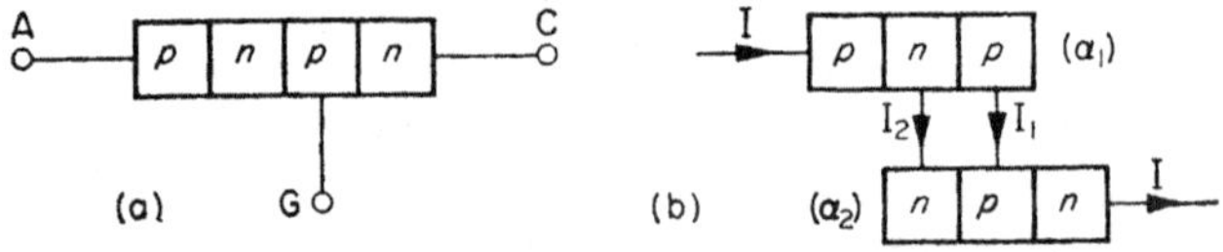

FIG. 5.20. Four-layer *pnpn* device. (a) Connections in the thyristor. (b) As complementary *pnp* and *npn* transistors.

5.9. The Silicon Controlled Rectifier

The four-layer, three-contact, *pnpn* device has a current/voltage characteristic such that it may be operated as a current controlled switch. The low power form of this device is known as a *pnpn* transistor of which a few germanium examples have been manufactured. The medium and high power *pnpn* silicon device is now widely used and is known as a controlled rectifier, or " thyristor ". One application of the *pnpn* transistor is considered in Chapter 9. In the thyristor (or S.C.R.) the possibility of conduction in the forward direction is controlled by an electrode, on one of the inner layers, known as a gate. This is shown simply in Fig. 5.20(a). Once conduction has started it cannot usually be stopped by means of the gate, but only by bringing the anode to cathode current below a critical value, known as the sustaining or " holding " current I_S. One version of the controlled rectifier, known as the " gate controlled switch ", is made in such a way that it can be turned off from the gate circuit, and this is useful in many medium power circuits. The thyristor has a reverse current/voltage characteristic similar to that of a silicon *pn* diode of a comparative size.

The principle of operation of the thyristor may perhaps be better understood by a consideration of Fig. 5.20(b). Here the four-layer device is represented as a pair of complementary *pnp* and *npn* transistors, with one *n* and one *p* region joined in common. The *pnp* transistor has a common base current gain of α_1 and the *npn* transistor a common base current gain of α_2. We will assume no gate current, such that the current flowing into the emitter of the *pnp* transistor is the same as that flowing from the emitter of the *npn* transistor. Then the collector currents are

$$I_1 = I_{1cbo} + \alpha_1 I,$$
$$I_2 = I_{2cbo} + \alpha_2 I,$$

where I_{1cbo} and I_{2cbo} are the leakage currents of the two transistors and

$$I = I_1 + I_2$$

or $$I = I_{1cbo} + I_{2cbo} + I(\alpha_1 + \alpha_2).$$

If we put $I_{cbo} = I_{1cbo} + I_{2cbo}$ = Combined leakage currents,

then $$I = \frac{I_{cbo}}{1-(\alpha_1+\alpha_2)}.$$

From this expression we can see that if $\alpha_1 + \alpha_2 \to 1$, then the total current becomes very large and the forward impedance of the S.C.R. tends to a very low value. This condition requires that the internal current gains (α_1 and α_2) of the S.C.R. should be controllable.

With very low values of current in a transistor-like device the current gain is small, but rises fairly rapidly as current is increased. Injection of current carriers into either base region (usually the *p* region in a British-made S.C.R.) increases the current flowing and the sum of the current gains quickly approaches unity. In a typical medium power *pnpn* device a gate current of 10 mA may control a main current of 25 A, with an anode to cathode volt drop of 1 V. With a high-power thyristor conducting 150 A, the volt drop will approach 2 V, to give a power dissipation of about 300 W.

The current/voltage characteristic is illustrated in Fig. 5.21. If the gate is left open-circuit, an increase of forward voltage initially results in a flow of leakage current. At some critical voltage, called " Break-over ", the current has a value of I_{BO}, and the characteristic curve breaks over into the low resistance, conducting, condition due to the reason outlined previously. Then to

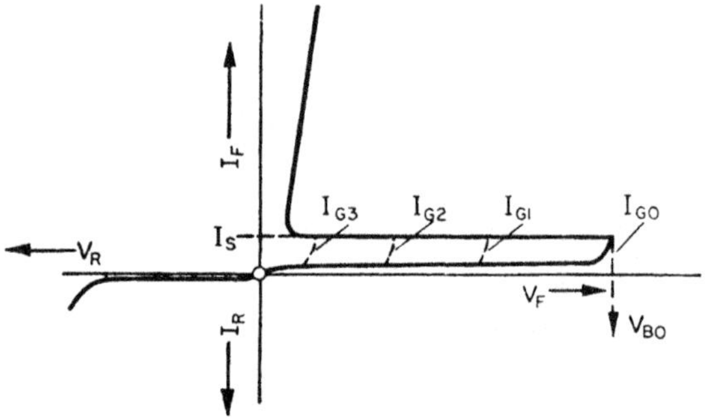

FIG. 5.21. Current/voltage characteristics of a thyristor.

return to the high resistance condition the main current must be reduced below the value I_S. With increasing values of gate current (such as I_{G1}, I_{G2} and I_{G3}) the forward curve breaks over at lower voltages. At a certain value of gate current the forward curve will be that of a single-junction diode. In addition, when current is flowing into the gate or base region, the thyristor acts like a transistor, with reverse voltage applied, so that the reverse current depends on the gate current flowing. The power dissipation may therefore be large under conditions of maximum reverse voltage. In practice the thyristor would not be turned on with zero gate current, as the large voltage required may cause permanent damage, and is operated as a switch, either fully conducting or fully blocked.

The switching time of the thyristor is short when compared with gas-filled or mechanical relays. The turn-on time is roughly 2 μsec, while the turn-off time is dependent on the external circuit. Turn-off cannot occur until the current has fallen below the sustaining value, and in an a.c. circuit this will be at the end of the half-cycle in which turn-off is initiated. After that point the minimum turn-off time is 10 μsec, but is actually dependent on the load current.

The thermal capacity of a thyristor is low, with a weight of only a few ounces in a volume of one cubic inch for a high power device. This means that the thyristor cannot withstand sustained overloads. High speed H.R.C. fuses are needed for protection, together with a series impedance, to give limited overload current. To prevent spurious firing due to voltage transients, which are usually present in high power equipment, screening of the control circuit is often essential.

5.10. Applications of the Thyristor

A simple circuit to demonstrate the use of a controlled rectifier in a half-wave application is in Fig. 5.22. With the component values as shown a 100 W lamp may be easily controlled. The disadvantage is that, with half-wave operation there is a noticeable flicker from the lamp. The output voltage from the phase-shifting network is applied to the thyristor gate, and this controls the time

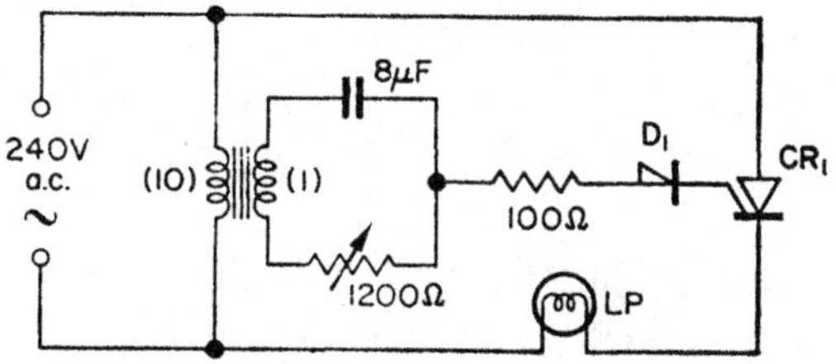

FIG. 5.22. Half-wave controlled rectification.

in the half-cycle at which the thyristor actually does conduct. This control of angle of conduction may be clearly seen if an oscilloscope is connected to observe the waveform across the load.

If a bridge circuit is employed, with a controlled rectifier in each of two arms, then control of d.c. power can be achieved.

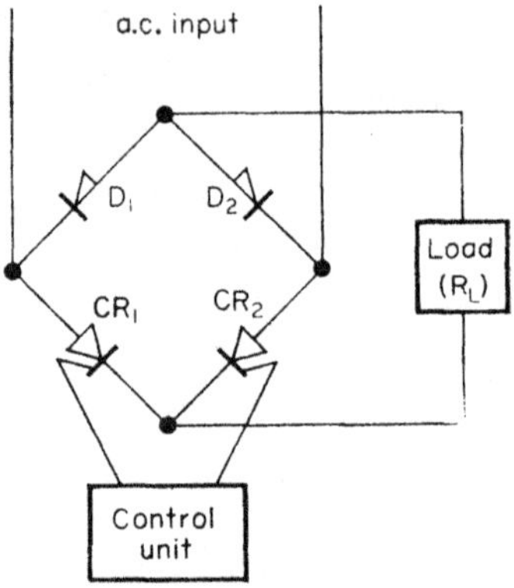

FIG. 5.23. D.C. power control with pulse firing.

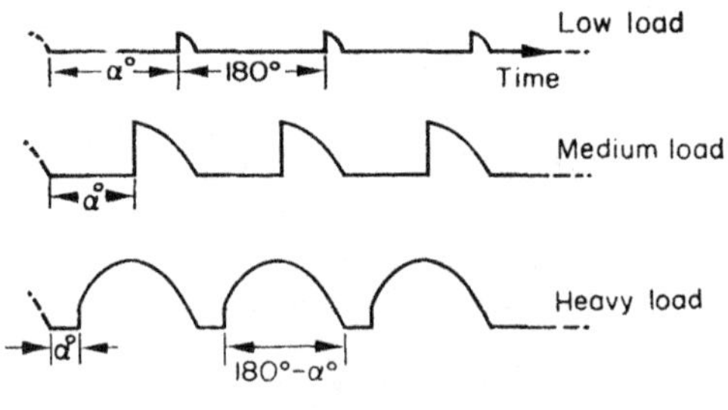

FIG. 5.24. Output waveforms for various firing angles.

This is illustrated in Fig. 5.23, where the firing point of the thyristor is controlled from a separate circuit. The control unit supplies a firing pulse train, with a variable mark : space ratio, and thus controls the firing angle, which is indicated as $\alpha°$ in Fig. 5.24.

The mean output power may be improved and the ripple percentage reduced, by the addition of a choke in series with the load as with any other rectifier circuit. If an inductance is joined in series with the load then a by-pass diode must be added across choke and load. This diode will then carry the direct current set up by discharge of the inductance. Without a low resistance path, this discharge current would flow through one diode and one thyristor in the bridge, and make it impossible to turn off the thyristor.

A d.c. supply may be inverted to a form of alternating current if the gates of two controlled rectifiers are fed with driving pulses

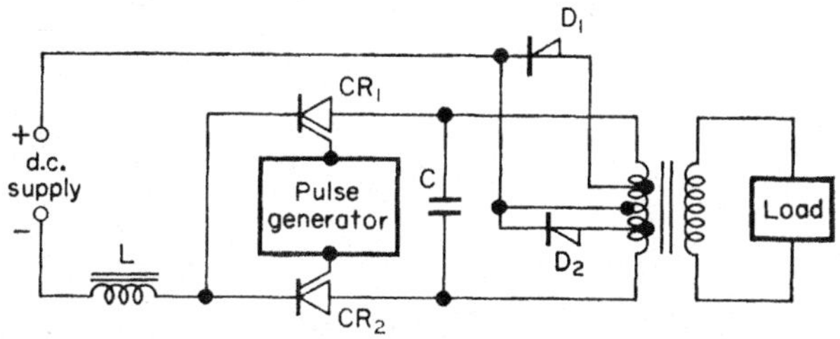

FIG. 5.25. Inverter circuit with two controlled rectifiers.

alternately at a given rate. The waveform may be made reasonably sinusoidal if a tuned circuit, with a Q factor of about 5, is on the output side.

The basic inverter circuit is Fig. 5.25. When CR_1 is ON, the d.c. supply is connected across one half of the transformer primary, and the voltage across the capacitor C is twice the supply voltage. Then when CR_2 is turned ON by a gate pulse, the capacitor is connected directly across CR_1, with a voltage in reverse (relative to CR_1), thus turning that rectifier OFF. The cycle is repeated as the pulse generator turns CR_1 ON and CR_2 OFF. The use of a gate controlled switch would simplify this circuit, since the control unit could provide turn-off as well as turn-on pulses and the turn-off capacitor would not be required. Turn-off of controlled rectifiers is one of the more difficult problems associated with this type of circuit.

The choke in Fig. 5.25 is in series with the supply source, which is usually a battery, and smooths the drain of current from the supply. The output waveform is basically a square wave modified by the presence of inductance and capacitance in the circuit. The output voltage tends to rise with a fall in the output current, but this is limited by the two clipping diodes connected to the quarter tappings on the output transformer.

Typical applications of inverters are to supply alternating current for mobile equipment where only direct current (say 12 or 28 V) exists. Operation at 400 c/s is common as a 50 c/s

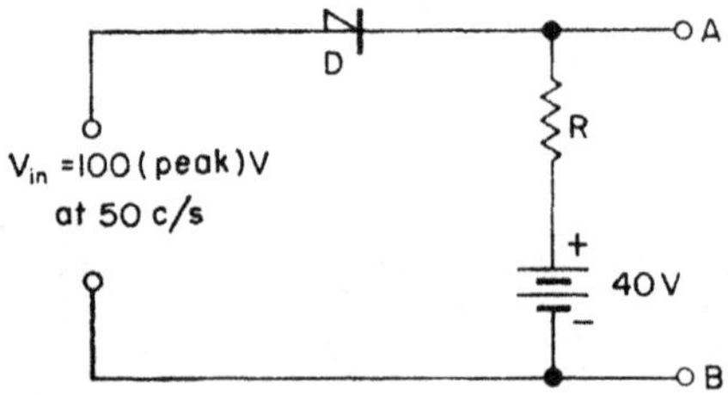

FIG. 5.26. Diode circuit for Q.5.1.

switching frequency would require a larger transformer. A three-phase inverter, with six controlled rectifiers, can give an output of 5 kW at 400 c/s with a true efficiency of 95%.

Questions for Chapter 5

Q.5.1. Deduce and sketch the waveform of V_{AB} at the output terminals *A* and *B* of Fig. 5.26, under the following conditions:

(a) With the circuit as shown.

(b) As in Fig. 5.26, but with the battery reversed.

(c) With both battery and diode reversed.

Resistance *R* is large with respect to the source resistance and the forward resistance of the diode. State clearly the steps of your deduction and show a voltage scale on the sketch.

Show how the waveforms would be modified if a resistance equal to *R* were put in the anode lead to the diode. (I.E.E., Dec. 1959)

Q.5.2. A half-wave diode rectifier feeds a load of resistance 10 kΩ, across which is connected a 1 μF capacitor. The supply is sinusoidal, of Frequency 50 Hz, and the diode has zero resistance when conducting, and infinite resistance when non-conducting. Determine the instants in the cycle when the diode conduction commences and ceases. (I.E.E., Oct. 1956)

Q.5.3. The germanium *P-N* junction is finding application as a power rectifier for both single and polyphase working. Explain briefly its principle of action discussing advantages and disadvantages compared with a mercury-arc rectifier. (I.E.E., Nov. 1958)

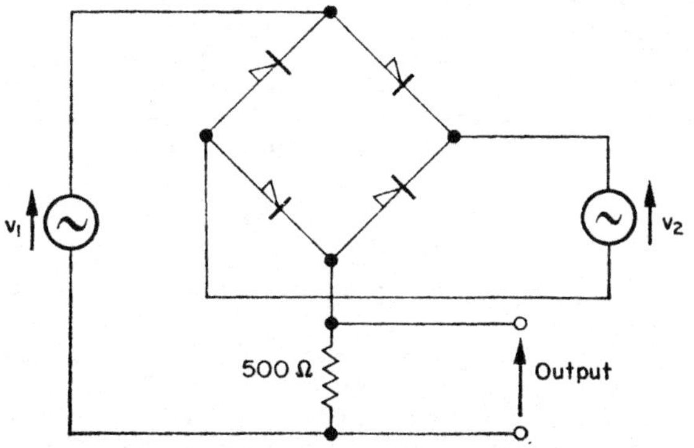

FIG. 5.27. Phase-sensitive rectifier circuit for Q.5.4.

Q.5.4. Explain briefly the purpose of a phase-sensitive rectifier. In the circuit of Fig. 5.27 the sine wave voltages v_1 and v_2 are in phase and the peak value of v_1 is 10 V. Each diode has a forward resistance of 500 Ω, and the reverse resistance of each may be taken as infinite. Draw on the same graph the waveform of the output voltage and of voltage v_1. Take the peak value of v_2 as

(a) 20 V, (b) 8 V and (c) 4 V.

In each case give numerical values for the peak output voltage, and state which diodes are conducting. (I.E.E., Nov. 1960)

Q.5.5. Explain the operation of a silicon junction diode used as a voltage reference source.

A zener diode and one resistor are joined in series to provide an 18 V stabilized output from a 22 V supply. If the load resistance is 150 Ω and the current through the stabilizer is 12 mA, determine the value of the series resistor and the power dissipated in it.

If the load resistance is increased to 200 Ω determine the current which then flows through the stabilizer diode. Assume operation of the diode is as an ideal switch. (H.N.C.)

Q.5.6. Describe briefly a method to determine the forward and reverse characteristics of low-power germanium and silicon diodes. Include in your description reasons both for the choice of apparatus and for the way it is connected. Comment on the possible causes of inaccuracy.

Indicate on suitable graphs (showing clearly the type of diode and the scales used) typical results of the experiment. Suggest an approximation for the relationship between applied voltage and the resistance of the diodes.

(H.N.D.)

Q.5.7. Explain briefly how a semiconductor junction diode may be used as a voltage stabilizer.

A silicon diode has a zener voltage of -6 V. Beyond that voltage this diode conducts in the reverse direction with a constant slope resistance of 6 Ω to a maximum current of 40 mA. Such a diode is joined in series with a ballast resistor of 120 Ω to stabilize the voltage across a resistive load. The load current may vary between 20 and 40 mA and the battery supply voltage from 13 to 11 V.

Find the stabilization ratio and the output resistance of this circuit.

(H.N.D.)

Q.5.8. Why are d.c. stabilizer circuits using gas-filled diodes not convenient for low-voltage applications?

Sketch the circuit of a simple stabilizer, using a silicon junction reference diode, and a power transistor as the series element. Explain the function of each component and indicate suitable component values for a circuit to supply 1 A at 6 V from a nominally 12 V supply (which may vary by ± 2 V).

(H.N.D.)

Q.5.9. Explain the operating principle of the *pnpn* type silicon controlled rectifier. Use a simple sketch of the device plus any current–voltage characteristics necessary to clarify the explanation.

Give one application of this rectifier, describing its operation with the aid of a basic circuit diagram. (H.N.C.)

CHAPTER 6

Voltage Amplifiers

THIS chapter deals with the application of transistors as voltage amplifiers. Typical circuits are analysed using load lines and equivalent circuits, and reference is made to some of the difficulties encountered at high frequencies.

6.1. Basic Amplifier Circuits

The three circuit configurations common (or grounded) base, common emitter and common collector are shown for a *pnp* transistor in Figs. 6.1(a), 6.1(b) and 6.1(c) and for an *npn* transistor in Figs. 6.1(d), 6.1(e) and 6.1(f). In each case the amplifying action depends on low power input circuits controlling high power output circuits. In the common base circuit, although the input and output currents are approximately equal, the low impedance emitter circuit absorbs far less power than is available at the high impedance collector. The low base current flowing into the common emitter circuit, where the impedance is a few thousand ohms gives rise to a much larger collector current flowing out of the high impedance output circuit. The common collector circuit with approximately equal input and output voltages, requires a low input current to control a much larger output current.

6.2. D.C. Biasing and Stabilization

Suitable operating potentials and currents may be obtained by the correct choice of supply voltage(s) and load and bias resistors. The simplest forms of supply circuit are shown in Fig. 6.1, but such circuits do not allow for the effects of leakage currents. As stated in Chapter 3, such leakage currents are very dependent upon temperature and although usually negligible compared with

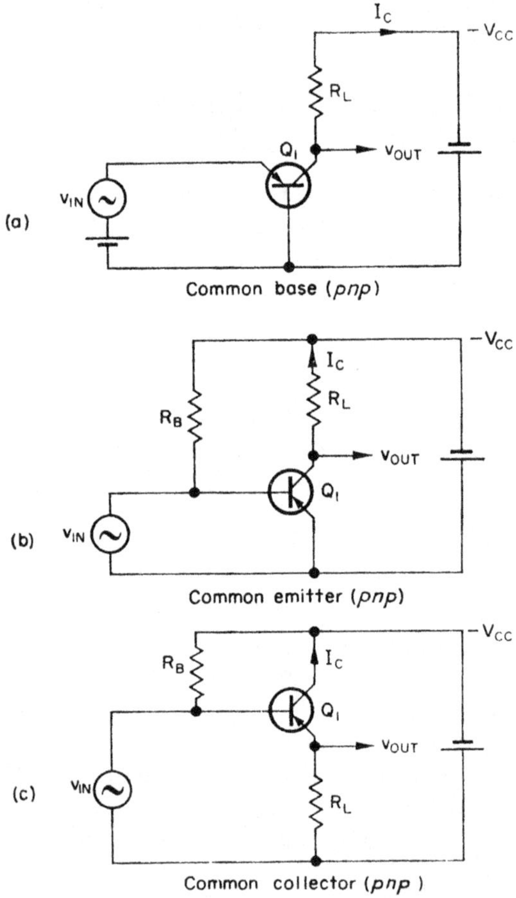

FIG. 6.1. Transistor circuit configurations.

the collector current, in the case of the grounded base circuit, are important for grounded emitter circuits. A typical value of the common base leakage current (I_{co}) for a low power audio frequency transistor is 8 μA at 25°C rising to 40 μA at 45°C while the corresponding values (I'_{co}) for the common emitter circuit are 110 and 550 μA.

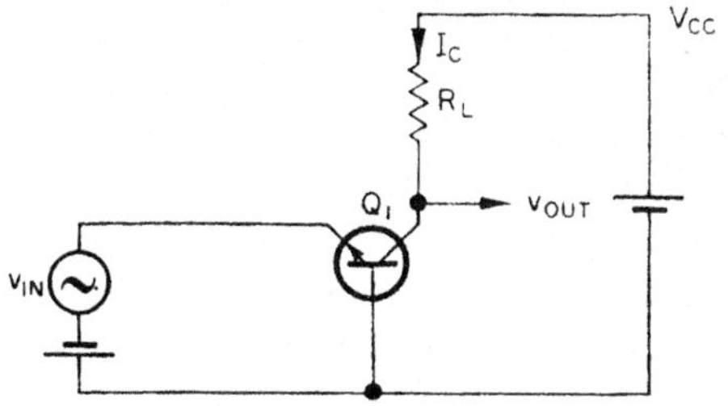

(d) Common base (*npn*)

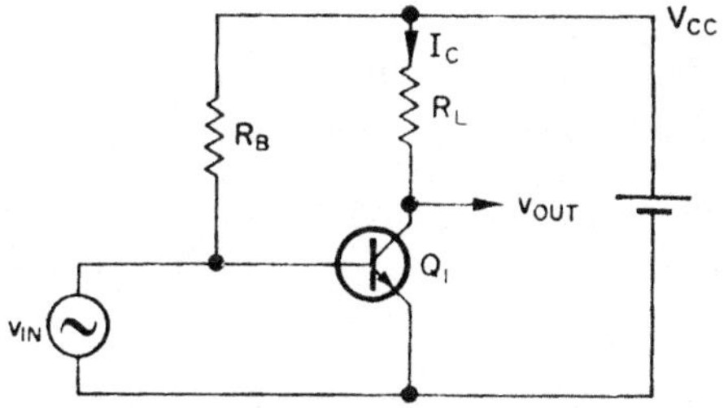

(e) Common emitter (*npn*)

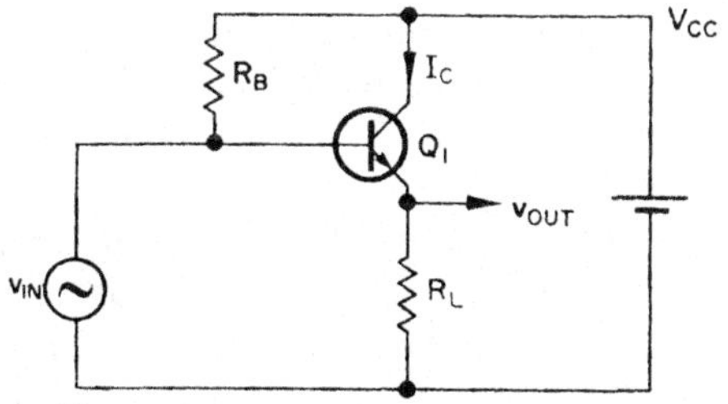

(f) Common collector (*npn*)

FIG. 6.1 (*cont.*)

As shown previously the d.c. value of the common emitter collector current is given by $I_c = \bar{\beta} I_B + I'_{co}$ and is dependent on I'_{co}. This shift of the operating point due to temperature change may give rise to clipping of the alternating output, bottoming of the transistor or, under worse conditions, thermal runaway. The latter effect occurs when an increase of leakage current raises the temperature of the transistor and gives rise to a further increase of leakage current. This cumulative effect may cause the destruction of the transistor.

Bias circuits should therefore be designed with due allowance for the effects of temperature, supply voltage variation and manufacturers' tolerances. Other electrical considerations are the current drain and amplifier input impedance, while cost and space limits the number of supplies and circuit components.

The degree of stabilization produced by any circuit is usually stated in terms of a factor of stability (K). The factor of stability is defined by the ratio:

$$K = \frac{\text{change in collector current with stabilization}}{\text{change in collector current without stabilization}}$$

and should be as low as possible. The three following sections refer to *pnp* circuits, but the treatment of *npn* circuits is obviously very similar.

6.3. Feedback Resistor Stabilization

The simplest form of stabilization circuit is shown in Fig. 6.2. I_E, I_B and I_C are the direct currents flowing through the transistor due to the applied voltage $-V_{cc}$, R_L is the load resistance and R_B the bias and stabilization resistance.

The operation of the circuit depends on d.c. feedback due to the resistance R_B. Any increase in I_C will cause a reduction in the collector voltage. This in turn will cause a reduction in the base emitter voltage V_{BE} and a decrease in I_B. A fall in I_B leads to a reduction in I_C and the original change in I_C is reduced.

Quantitatively, V_{BE} is very low and the following equations apply to the circuit:

$$V_{cc} = (I_B + I_c)R_L + I_B R_B, \quad \text{since } V_{BE} \text{ is very low,}$$

$$I_c = \beta I_B + I'_{co},$$

which leads to

$$I_c\left(1+\frac{\beta R_L}{R_L+R_B}\right) = \frac{\beta V_{cc}}{R_L+R_B}+I'_{co}.$$

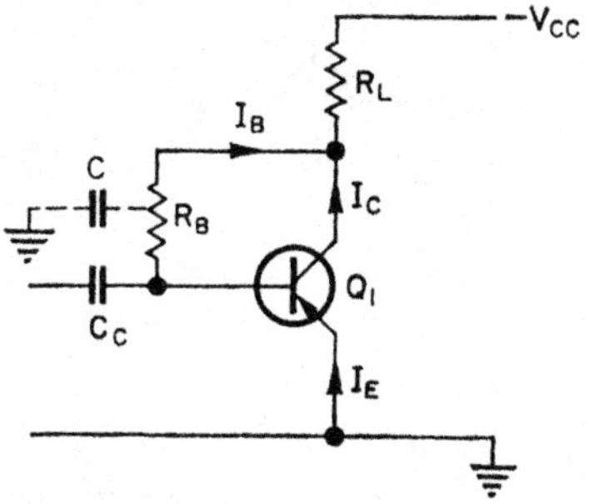

FIG. 6.2. Feedback resistor stabilization.

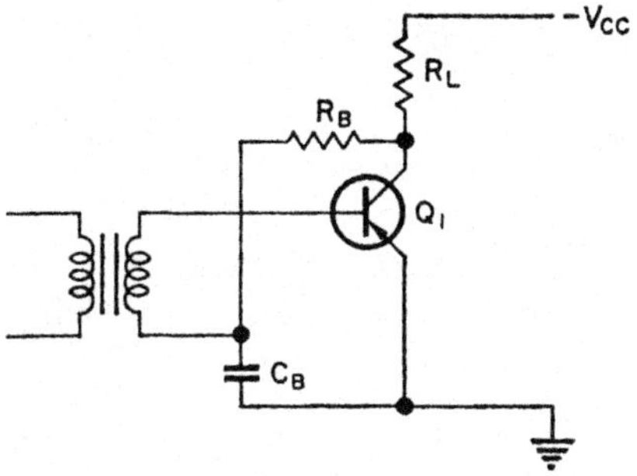

FIG. 6.3. Feedback resistor stabilization with transformer coupling.

Hence differentiating partially with respect to I'_{co},

$$\frac{\partial I_c}{\partial I'_{co}} = \frac{1}{1+\beta R_L/(R_L+R_B)}.$$

Since $\partial I'_{co}$ is the change in collector current without stabilization this expression gives the factor of stability K. For optimum stability R_L should be much greater than R_B which implies a high value of V_{cc} and a high standing value of the collector current I_C. In practice R_B is decided by bias requirements and not stability considerations and is usually very much greater than R_L.

A disadvantage of this circuit is that a.c. feedback is also present but this may be nullified by a decoupling capacitor C as shown in Fig. 6.2. The corresponding circuit for a transformer-coupled amplifier is shown in Fig. 6.3.

The expression for I_c may also be differentiated partially with respect to V_{cc} and $\bar{\beta}$ giving

$$\frac{\partial I_c}{\partial V_{cc}} = \frac{\bar{\beta}}{R_B + R_L(1+\bar{\beta})} \quad \text{and} \quad \frac{\partial I_c}{\partial \bar{\beta}} = \frac{V_{cc} - I_c R_L}{R_B + R_L(1+\bar{\beta})}.$$

These relations show the effects of supply voltage and current gain on the stability of the collector current, but, as before, R_B is decided by bias considerations and not optimum stability.

6.4. Base Resistor and Emitter Bias Stabilization

This circuit, shown in Fig. 6.4, depends for its operation on the feedback resistor R_E which is decoupled for a.c. signals. Any increase in I_c means an increase in I_E which increases the volt drop across R_E. The consequent reduction in V_{BE} reduces I_B and hence reduces I_c counteracting the original change.

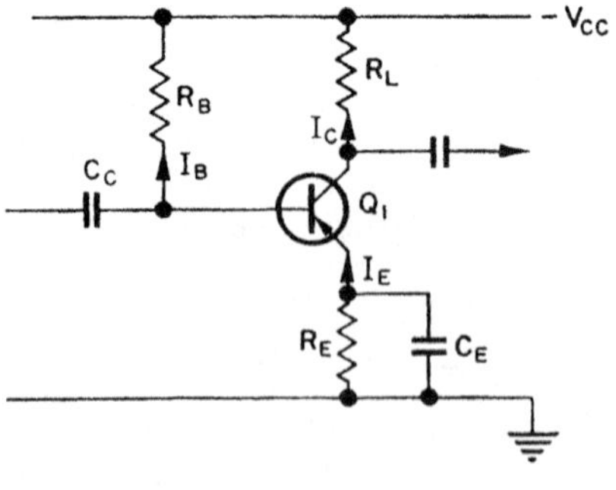

FIG. 6.4. Base resistor and emitter bias stabilization.

If we ignore the low voltage drop between base and emitter the following d.c. equations apply

$$V_{cc} = I_E R_E + I_B R_B,$$
$$I_c = \beta I_B + I'_{co},$$
$$I_c = I_E - I_B.$$

Eliminating I_E and I_B gives

$$I_c\left(1+\frac{\beta R_E}{R_E+R_B}\right) = \frac{\beta V_{cc}}{R_E+R_B}+I'_{co}.$$

Differentiating gives the factor of stability

$$K = \frac{\partial I_c}{\partial I'_{co}} = \frac{1}{1+\beta R_E/(R_E+R_B)}.$$

K decreases as R_E is increased but in practice both R_E and R_B are decided by bias rather than stability conditions. The capacitor C_E eliminates a.c. feedback and is usually about 50 μF.

Further differentiating gives the effect of changes in V_{cc} and β on the collector current.

$$\frac{\partial I_c}{\partial V_{cc}} = \frac{\beta}{R_E(1+\beta)+R_B},$$

$$\frac{\partial I_c}{\partial \beta} = \frac{V_{cc}-I_c R_E}{R_E(1+\beta)+R_B}.$$

6.5. Potential Divider and Emitter Bias Stabilization

This method is the most widely used and the stabilizing circuit is shown in Fig. 6.5. As in the previous case an increase in I_c leads to an increase in I_E and a larger volt drop across R_E.

Since the base potential is approximately constant V_{BE} is reduced, I_B falls and I_c is reduced, thus tending to counteract the original change. If we ignore the low value of V_{BE} as before,

$$V_{cc} = I_1(R_1+R_2)+I_B R_1,$$
$$I_1 R_2 = I_E R_E = (I_B+I_c)R_E,$$
$$I_c = \beta I_B + I'_{co}.$$

The first two equations give

$$V_{cc} = \frac{(I_B+I_c)}{R_2} R_E(R_1+R_2)+I_B R_1$$

or

$$\frac{V_{cc}}{R_1} = (I_B+I_c)\frac{R_E}{R_B}+I_B, \quad \text{where } R_B = \frac{R_1 R_2}{R_1+R_2}.$$

Hence

$$I_c\left(1+\frac{\beta R_E}{R_E+R_B}\right) = \frac{\beta V_{cc} R_B}{R_1(R_E+R_B)}+I'_{co}.$$

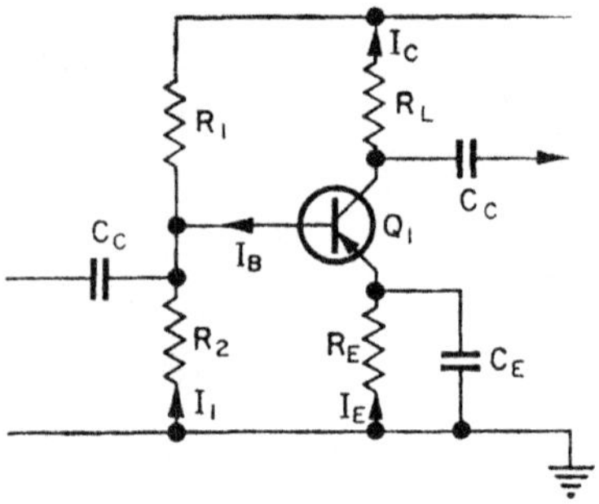

FIG. 6.5. Potential divider and emitter bias stabilization.

Differentiating partially,

$$K = \frac{\partial I_c}{\partial I'_{co}} = \frac{1}{1+\beta R_E/(R_E+R_B)}.$$

R_B $(= R_1R_2/(R_1+R_2))$ may be made comparable with R_E while still maintaining the ratio R_1/R_2 for correct bias, and in this case bias potentials do not affect the choice of stabilizing components. R_1 and R_2 for this circuit are limited by the d.c. drain and the shunting of the alternating input signal. Typically, R_1 is about ten times R_2, and R_2 about ten times R_E.

The effect of variation of supply voltage and current gain is obtained as before by differentiating.
i.e.

$$\frac{\partial I_c}{\partial V_{cc}} = \frac{R_B}{R_1} \cdot \frac{\bar{\beta}}{[R_B+R_E(1+\bar{\beta})]},$$

$$\frac{\partial I_c}{\partial \bar{\beta}} = \frac{V_{cc}R_B - I_c R_1 R_E}{R_1[R_B+R_E(1+\bar{\beta})]}.$$

6.6. Analysis and Design of Circuits Using Load Lines

To illustrate the principles involved in analysis and design two cases will be considered, the analysis of a common base amplifier and the design of a common emitter amplifier. Solutions using load lines involve the superposition of lines representing the variation of collector voltage with collector current on static characteristics.

To find the operating point the d.c. load line is used. That is,

$$V_{CE} \text{ or } V_{CB} = -V_{CC} + I_c R_L$$

for the *pnp* circuits given in Fig. 6.1(a) and 6.1(b), and

$$V_{CE} \text{ or } V_{CB} = V_{CC} - I_C R_L$$

for the *npn* circuits given in Fig. 6.1(d) and 6.1(e).

Where this line intersects the static characteristic for the standing value of emitter (or base) current gives the operating collector voltage and current. The slope of the d.c. load line is

$$\frac{\delta I_c}{\delta V_C} = \pm \frac{1}{R_L}$$

depending on whether the circuit is for *pnp* or *npn* transistors.

For the alternating current the collector load is often different from that for the direct current. The a.c. load line is represented by a straight line of slope $\pm 1/Z_L$ passing through the d.c. operating point, where Z_L is the a.c. load at the collector. The collector voltage and current swing for a given change in emitter (or base) current may then be determined. The procedure is illustrated in the examples given in Sections 6.6.1, 6.6.2 and 6.9.1.

6.6.1. *Analysis of a common base amplifier*

Consider the simple circuit shown in Fig. 6.6, the transistor itself having the static characteristics shown in Fig. 6.7(a) and (b.)

The d.c. input resistance is very low and the standing emitter current is very nearly 2 V/4 kΩ, i.e. 0·5 mA. The a.c. input

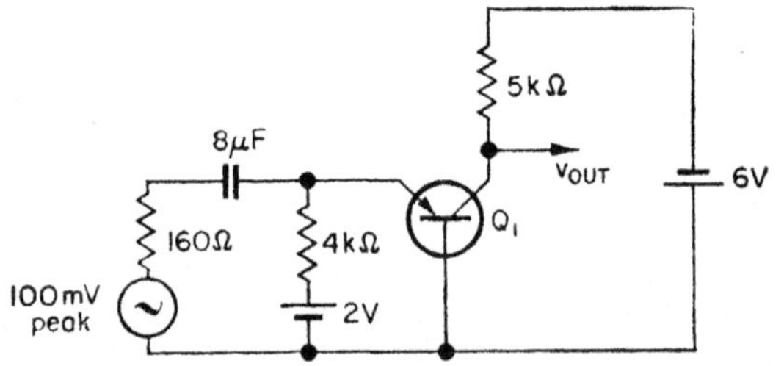

FIG. 6.6. Common base amplifier.

resistance at this value of emitter current is 8 mV/0·2 mA = 40 Ω giving a peak value of the alternating input current of 100 mV/200 = 0·5 mA.

The load line gives the variation of I_c with collector voltage, i.e. $V_{cB} = -6+5I_c$ (I_c in milliamps) and this may be plotted on the same axes as the static characteristics. Where this load line intersects the characteristics for $I_E = 0{\cdot}5$ mA is the operating collector current and potential (i.e. −3·8 V, 0·45 mA) and the product of these (1·71 mW) gives the power at the collector.

Since the variation in emitter current is 0·5 mA peak, the projections of the intersections of the load line and the static characteristics for emitter currents of 0 and 1 mA give the swing in output voltage (4·4 V peak to peak). The overall voltage gain is then $4{\cdot}4/200 \times 10^{-3} = 22$ times.

The current swing is 0·87 mA peak to peak so that the amount of d.c. power turned into a.c. power at the collector is $V_{\text{r.m.s.}} I_{\text{r.m.s.}}$, i.e. $(0{\cdot}87/2\sqrt{2}) \times (4{\cdot}4/2\sqrt{2}) = 0{\cdot}48$ mW. The difference between this and 1·71 mW is the power that will be dissipated at the collector (i.e. 1·23 mW). This is well within the rating of the transistor which is probably about 25 mW.

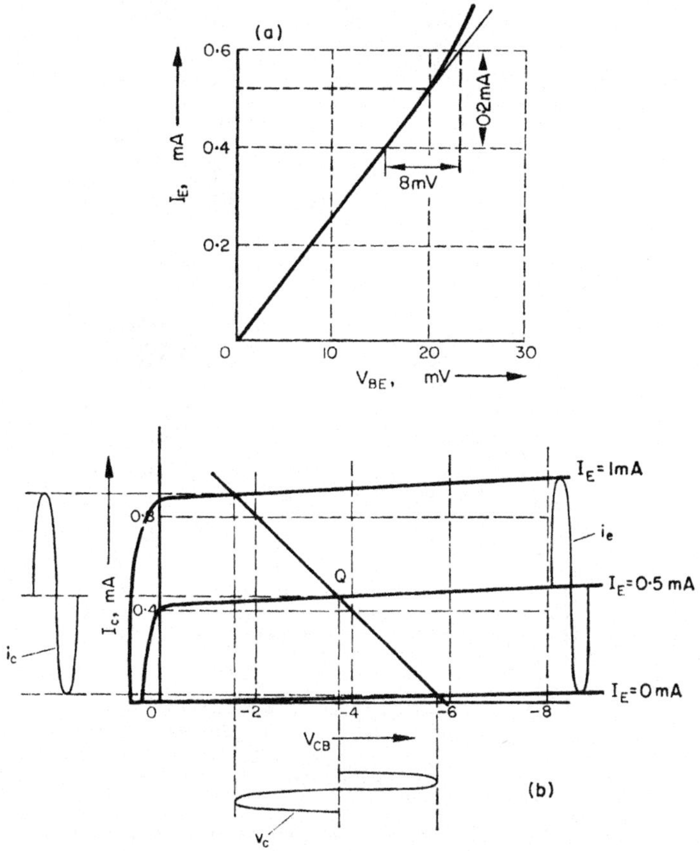

FIG. 6.7. Characteristics of transistor used in Fig. 6.6.

6.6.2. *Design of a common emitter amplifier*

For the purposes of this illustration assume that a low level a.f. amplifier operating into a high impedance load is required.

Assume a 9 V supply is available, the source impedance is 1 kΩ, and the transistor to be used has the characteristics shown in Fig. 6.8(a) and (b).

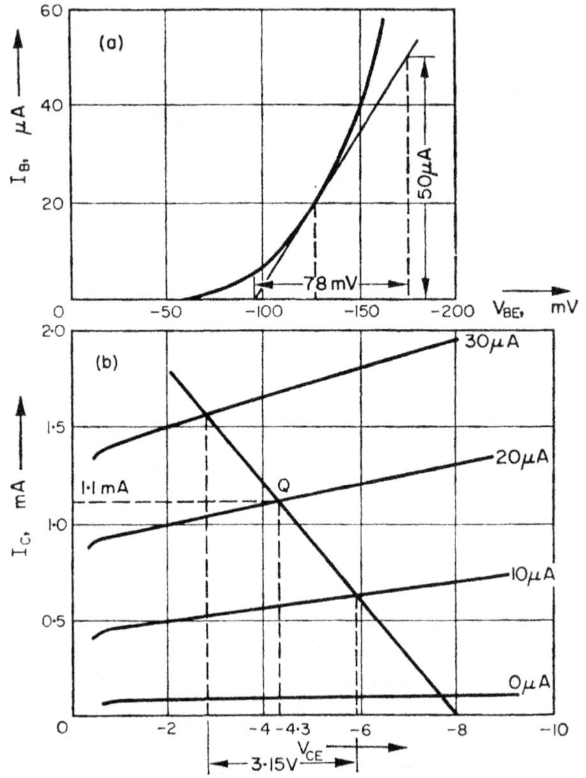

FIG. 6.8. Common emitter characteristics used for design of simple amplifier.

Consider the circuit shown in Fig. 6.9. A suitable direct voltage drop across the bias resistor R_E would be about 1 V, since this would leave 8 V to be dropped across the transistor and load resistance R_L.

The load line is then represented by

$$V_{CE} = -8+I_c R_L.$$

By inspection, a suitable load line is shown in Fig. 6.8(b) and this corresponds to a load resistance R_L of 5 V/1·5 mA = 3·3 kΩ with quiescent values of collector–emitter voltage V_c = 4·3 V and collector current I_c = 1·1 mA at a base current of 20 μA.

The potential divider $R_1 R_2$ should not lower the a.c. input impedance appreciably, but should carry a direct current considerably greater than the d.c. base current. Hence for a base

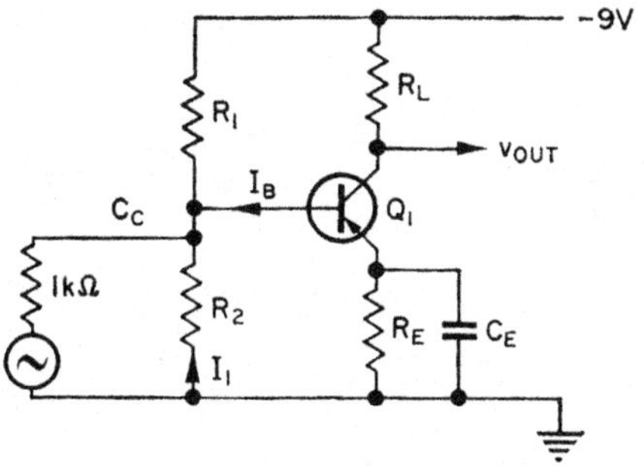

FIG. 6.9. Common emitter amplifier.

current of 20 μA, if R_2 = 7·5 kΩ, I_1 = 1·125 V/7·5 kΩ = 150 μA and R_1 is 7·875 V/170 μA = 47 kΩ.

These values are suitable and R_B, which is R_1 in parallel with R_2, is 6·5 kΩ. R_E must then be 1 V/1·1 mA = 900 Ω.

From the characteristics $\bar{\beta}$ = 1·1 mA/20 μA = 55, so that the factor of stability $K = 1/[1+(55\times 900/7400)] = 0{\cdot}13$ which should allow satisfactory operation over a reasonable range of temperature.

The a.c. input impedance at a base current of 20 μA is 78 mV/50 μA, i.e. 1·56 kΩ, and a base current swing of 20 μA requires an alternating base voltage of 31·2 mV. This voltage also appears across R_B so that the total current required from the source is 20+(31·2 mV/6·5 kΩ) = 24·8 μA.

The source voltage must then be $24 \cdot 8 + 31 \cdot 2 = 56$ mV and gives rise to an output voltage of 3·15 V, the latter value being obtained from the load line. The voltage gain is $3 \cdot 15/56 \times 10^{-3} = 56$ times.

If this is a higher value than required, a small non-bypassed resistance in the emitter lead will introduce negative feedback, reducing the gain and enhancing the performance of the stage.

The power dissipated at the collector with no alternating input is $1 \cdot 1 \times 4 \cdot 3 = 4 \cdot 7$ mW, which should be well within the rating of the transistor.

The capacitors should have a low reactance at the lowest frequencies and a typical value of C_c is 8 μF, while the emitter decoupling capacitor would be of the order of 50 μF.

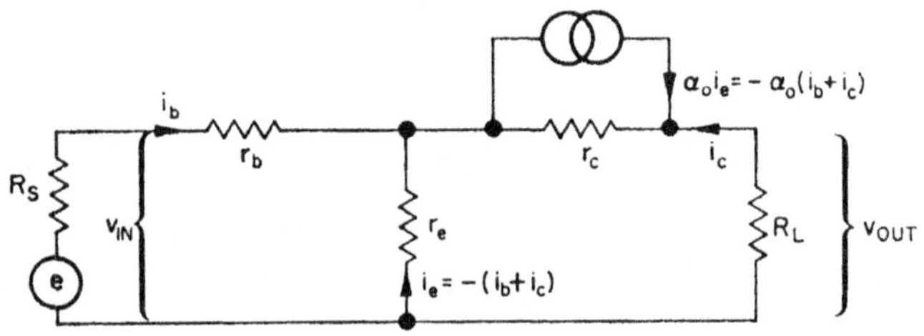

FIG. 6.10. Equivalent circuit of common emitter amplifier.

The circuit developed from the static characteristics indicates the sort of considerations which should be made in the design of amplifiers. If there is any choice of transistor, the main points to be considered are current gain, leakage current and power supplies required.

6.7. Analysis of Circuits Using the *T* Equivalent Circuit

When using the *T* equivalent circuit, it must be remembered that it only applies for small signal a.c. parameters and gives no indication of the d.c. supplies required. The parameters, however, depend upon operating direct currents and potentials and in general this method is much more useful for analytical rather than design purposes. The following paragraphs illustrate the use of the T equivalent circuit, algebraic results being developed for the grounded emitter circuit and arithmetic values being calculated for the grounded collector.

6.7.1. *Analysis of the common emitter circuit*

The simplest form of grounded emitter circuit is shown in Fig. 6.1 and may be represented by the equivalent circuit in Fig. 4.9. The complete equivalent circuit including the source e.m.f. and source and load resistances is shown in Fig. 6.10. The currents flowing into the transistor are conventionally shown as positive and R_B is assumed high enough to be ignored.

The equations for the circuit shown are

$$e = i_b(R_S+r_b+r_e)+i_c r_e \tag{1}$$

$$0 = i_c(R_L+r_c+r_e)+i_b r_e-\alpha_0(i_b+i_e)r_c$$

$$= i_c[R_L+r_c(1-\alpha_0)+r_e]+i_b(r_e-\alpha_0 r_c). \tag{2}$$

In practice r_e is small enough to be neglected in comparison with R_L and $r_c(1-\alpha_0)$ so that eqn. (2) gives

$$i_c[R_L+r_c(1-\alpha_0)] = i_b\alpha_0 r_c. \tag{3}$$

Eliminating i_c,

$$e = i_b(R_S+r_b+r_e)+\frac{i_b\alpha_0 r_c r_e}{R_L+r_c(1-\alpha_0)}.$$

If R_S is put equal to zero,

$$Z_{IN} = \frac{v_{IN}}{i_b} = r_b+r_e+r_e\left[\frac{\alpha_0}{(1-\alpha_0)+R_L/r_c}\right].$$

If R_L is very much less than r_c,

$$Z_{IN} = r_b+r_e/(1-\alpha_0).$$

Equation (3) gives the current gain:

$$\frac{i_c}{i_b} = \frac{\alpha_0 r_c}{R_L+r_c(1-\alpha_0)} = \frac{\alpha_0}{R_L/r_c+(1-\alpha_0)}.$$

If R_L is much less than r_c,

$$\frac{i_c}{i_b} = \frac{\alpha_0}{1-\alpha_0} = \beta_0.$$

The overall voltage gain

$$= \frac{v_{\text{OUT}}}{e} = -\frac{i_c R_L}{i_b(R_S+Z_{\text{IN}})}$$

$$= -\frac{\beta_0 R_L}{R_S+r_b+r_e/(1-\alpha_0)} \quad \text{(very nearly)}$$

$$= -\frac{\alpha_0 R_L}{(R_S+r_b)(1-\alpha_0)+r_e}.$$

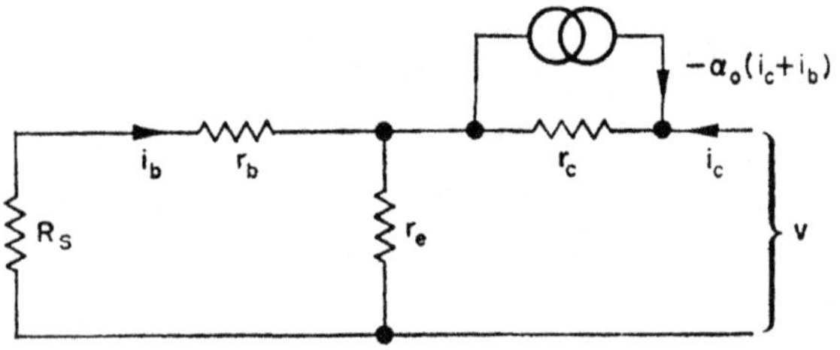

FIG. 6.11. Equivalent circuit of common emitter amplifier (output impedance).

The voltage gain of the transistor itself may be obtained by putting $R_S = 0$, i.e.

$$\text{Voltage gain} = \frac{v_{\text{OUT}}}{v_{\text{IN}}} = \frac{\alpha_0 R_L}{r_e+r_b(1-\alpha_0)}.$$

In this equation the negative sign has been omitted since it only indicates a 180° phase reversal.

The output impedance may be obtained by applying a voltage to the output terminals and short-circuiting the input e.m.f. as shown in Fig. 6.11.

The equations for this circuit are

$$v = i_c(r_c+r_e)-\alpha_0(i_c+i_b)r_c+i_b r_e$$

$$= i_c[r_e+r_c(1-\alpha_0)]+i_b[r_e-\alpha_0 r_c]$$

$$0 = i_b(R_S+r_b+r_e)+i_c r_e.$$

Eliminating i_b,

$$v = i_c[r_e+r_c(1-\alpha_0)]-\frac{r_e(r_e-\alpha_0 r_c)}{R_S+r_b+r_e}.$$

r_e may be neglected in comparison with $r_c(1-\alpha_0)$. Therefore

$$Z_{\text{OUT}} = \frac{v}{i_c} = r_c(1-\alpha_0)+\frac{\alpha_0 r_c r_e}{R_S+r_b+r_e}.$$

Typical values for an alloy junction transistor are 25 Ω for r_e, 1000 Ω for r_b, 1 MΩ for r_c and 0·98 for α_0. A source resistance of 1000 Ω and a load resistance of 5 kΩ then gives

Input impedance = 2250 Ω.

Current gain = 39.

Voltage gain of transistor itself = 109.

Output impedance = 32 kΩ.

In practice it is usually easier to substitute the values for r_e, r_b, r_c, α_0, R_S and R_L initially, thus giving equations that may be solved fairly easily. This procedure is illustrated in the following section.

6.7.2. *Analysis of a common collector circuit*

Assuming the transistor used in the common collector circuit shown in Fig. 6.12 has the same parameters as in the previous case, i.e. $r_e = 25$ Ω, $r_b = 1000$ Ω, $r_c = 1$ MΩ, $\alpha_0 = 0{\cdot}98$, the equivalent circuit is as shown in Fig. 6.13. As in the previous case R_B is assumed high enough and X_{C_c} low enough to be ignored.

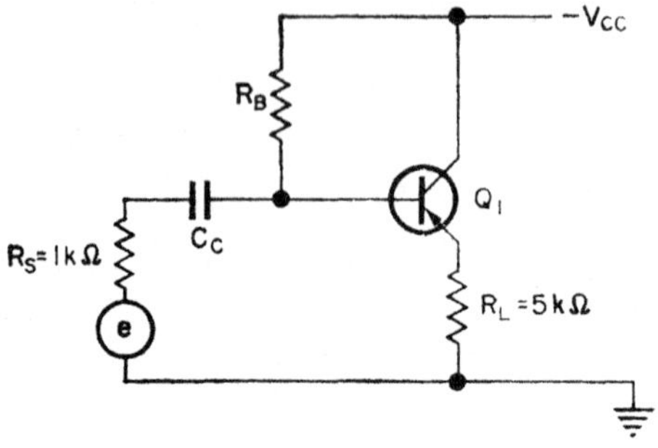

FIG. 6.12. Common collector amplifier.

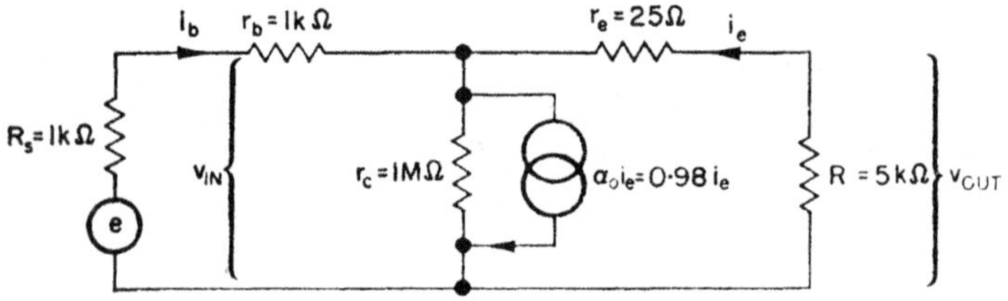

FIG. 6.13. Equivalent circuit of Fig. 6.12.

The equations for Fig. 6.13 are

$$e = i_b\, 1{\cdot}002\times 10^6 + 10^6(1-0{\cdot}98)i_e$$

$$= i_b \times 1{\cdot}002\times 10^6 + 2i_e\times 10^4 \tag{4}$$

$$0 = i_e\times 1{\cdot}005025\times 10^6 - 0{\cdot}98 i_e\times 10^6 + i_b\times 10^6$$

$$= 2{\cdot}5025\times 10^4 i_e + i_b\times 10^6.$$

Current gain

$$\frac{i_e}{i_b} = -\frac{10^6}{2{\cdot}5025\times 10^4} = 40 \text{ times (numerically).} \tag{5}$$

Eliminating i_e between eqns. (4) and (5),

$$e = i_b \times 1{\cdot}002 \times 10^6 - 80 i_b \times 10^4$$

$$\frac{e}{i_b} = 202\ \text{k}\Omega$$

$$Z_{IN} = \frac{v_{IN}}{i_b} = 202-1 = 201\ \text{k}\Omega.$$

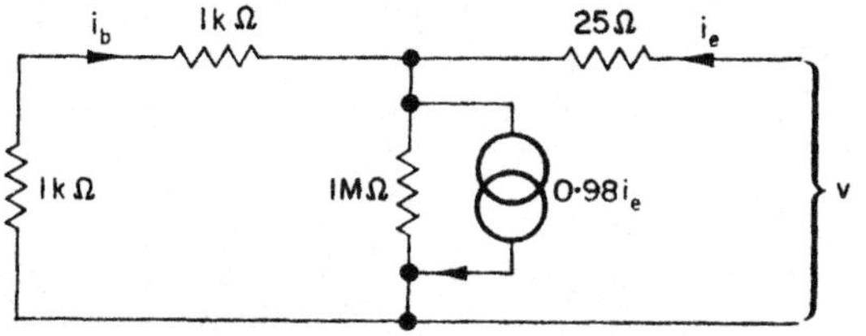

FIG. 6.14. Equivalent circuit of common collector amplifier (output impedance).

Overall voltage gain

$$= \frac{v_{OUT}}{e} = -\frac{i_e R_L}{i_b(R_S+Z_{IN})} = 0{\cdot}99.$$

Voltage gain of transistor itself

$$= \frac{40 \times 5000}{201000} = 0{\cdot}995.$$

To determine the output impedance apply a voltage v to the output and short-circuit the input e.m.f. as shown in Fig. 6.14.

The corresponding equations are

$$v = i_e(25 + {\cdot}02 \times 10^6) + i_b \times 10^6$$

$$0 = i_b(1{\cdot}002 \times 10^6) + i_e \times {\cdot}02 \times 10^6.$$

From these equations

$$\frac{i_b}{i_e} = -\frac{2\times 10^4}{1{\cdot}002\times 10^6} = -\frac{1}{50{\cdot}1}$$

Therefore $$v = i_e \times 20{\cdot}025\times 10^3 - 1/50{\cdot}1 \times i_e \times 10^6$$

i.e. $$Z_{\mathrm{OUT}} = \frac{v}{i_e} = 65\ \Omega.$$

TABLE 6.1. COMPARISON OF AMPLIFIER CIRCUIT PARAMETERS

	Common base	Common emitter	Common collector
Z_{IN}	$r_e + r_b(1-\alpha_0)$ (45 Ω)	$r_b + \frac{r_e}{1-\alpha_0}$ (2250 Ω)	$r_b + \frac{R_L}{R_L/r_c + (1-\alpha_0)}$ (201 kΩ)
Z_{OUT}	$r_c - \frac{\alpha_0 r_c r_b}{r_e + R_S + r_b}$ (516 kΩ)	$r_c(1-\alpha_0) + \frac{\alpha_0 r_c r_e}{R_S + r_b + r_e}$ (32 kΩ)	$r_e + (R_S + r_b)(1-\alpha_0)$ (65 Ω)
Current gain	$-\frac{\alpha_0}{1 + R_L/r_c}$ (0·98)	$\frac{\alpha_0}{R_L/r_c + (1-\alpha_0)}$ (39)	$-\frac{1}{(1-\alpha_0) + R_L/r_c}$ (40)
Voltage gain	$\frac{\alpha_0 R_L}{r_e + r_b(1-\alpha_0)}$ (109)	$-\frac{\alpha_0 R_L}{r_e + r_b(1-\alpha_0)}$ (109)	$\frac{1}{1 + (r_b/R_L)(1-\alpha_0)}$ (0·995)
Power gain	(107)	(4251)	(40)

6.8. Comparison of Transistor Circuits

Transistor circuits may be analysed by the methods described in the previous sections and the results for common base, emitter and collector are summarized in Table 6.1. The figures in brackets indicate the values for the assumed *T* equivalent circuit with $r_e = 25\ \Omega$, $r_b = 1000\ \Omega$, $r_c = 1\ \text{M}\Omega$, $\alpha_0 = 0{\cdot}98$, $R_S = 1\ \text{k}\Omega$ and $R_L = 5\ \text{k}\Omega$.

The common emitter and common collector circuits give high current gains while common base and common emitter give high voltage gains. The highest power gain is obtained using the common emitter circuit and this configuration is usually used in amplifier circuits. The high input and low output impedance of the common collector (or emitter follower) circuit make it useful for buffer amplifiers or probe units. The low input and high

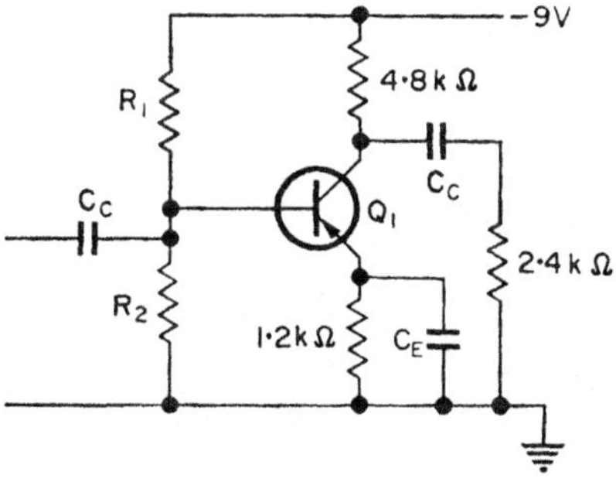

FIG. 6.15. R.C. coupled common emitter amplifier.

output impedances of the common base circuit render it unsuitable for cascade connection but common emitter stages may be connected in cascade and multi-stage, high gain amplifiers use this configuration. A common collector stage may be fed into a common base stage without much mismatch and this connection is sometimes used.

6.9. R.C. Coupling of Transistor Stages

Connection between transistor stages may be accomplished by a variety of methods, but the most popular at low and medium frequencies is R.C. coupling. It requires less space than L.C. or transformer coupling and has a better frequency response. At high frequencies, where only a narrow band is required to be amplified, tuned circuits with capacitive or inductive coupling are used.

6.9.1. *Load line of R.C. coupled transistor stage*

Assume Fig. 6.15 represents one stage of an amplifier, the input current of which is 10 μA peak. Let the output characteristics of the transistor be as shown in Fig. 6.16 and assume the bias components give a standing base current of 20 μA.

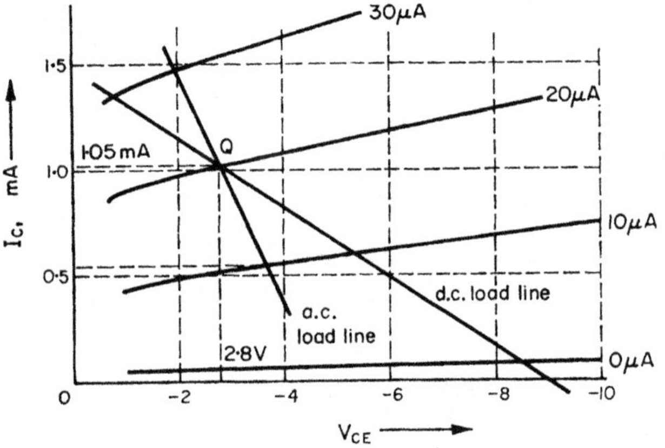

FIG. 6.16. Characteristics of transistor used in Fig. 6.15.

The operating point may be determined by drawing the d.c. load line. This gives the variation in collector voltage for different direct currents flowing through the transistor, and since the total resistance between collector and emitter is 6 kΩ, is represented by the equation:

$$V_{CE} = -9 + I_c 6 \quad (I_c \text{ in milliamps}).$$

This assumes the collector and emitter currents are equal, but the error in ignoring the base current is small. Where this load line intersects the 20 μA characteristics gives the operating point of −2·8 V, 1·05 mA.

For alternating inputs the load impedance is 4·8 kΩ in parallel with 2·4 kΩ if we assume the coupling capacitor has negligible reactance. Hence if we draw a line of slope 1·6 kΩ through the operating point the effect of alternating inputs may be estimated. This is the a.c. load line and shows a collector current swing 0·95 mA for an input swing of 20 μA, and two-thirds of this (i.e. 0·64 mA) flows through the 2·4 kΩ resistance which represents the input impedance for the next stage. A swing of 0·64 mA peak to

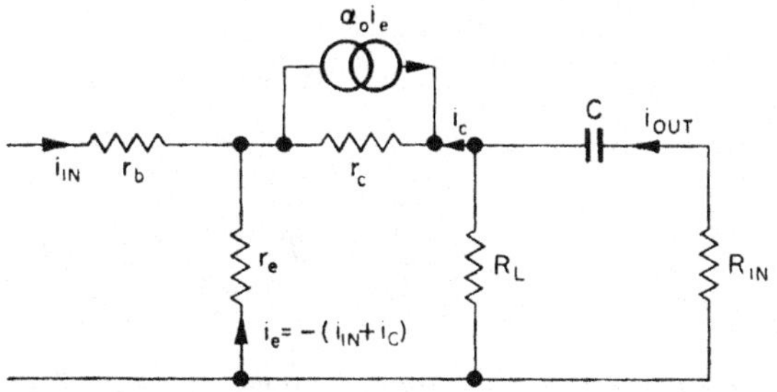

FIG. 6.17. L.F. equivalent circuit of Fig. 6.15.

peak at the base of the following transistor would develop appreciable collector current variation and enable a high overall voltage gain to be obtained. The output current at the collector of the stage shows slight harmonic distortion (about 5%) but this would be tolerable.

6.9.2. *Frequency response of a R.C. coupled stage*

The low frequency response of such a stage may be obtained by reference to the simple *T* equivalent circuit and is shown in Fig. 6.17. If Z is the impedance of R_L in parallel with C and the input impedance of the next stage,

$$i_c(Z+r_c+r_e)-\alpha_0(i_{IN}+i_c)+i_{IN}r_e = 0.$$

Therefore

$$\frac{i_c}{i_{IN}} = \frac{\alpha_0 r_c}{Z+r_c(1-\alpha_0)} \quad \text{(very nearly, since } r_e \text{ is small).}$$

Therefore

$$\frac{i_{\text{OUT}}}{i_{\text{IN}}} = \frac{i_{\text{OUT}}}{i_c} \times \frac{i_c}{i_{\text{IN}}} = \frac{Z}{R_{\text{IN}}+(1/j\omega c)} \times \frac{\alpha_0 r_c}{Z+r_c(1-\alpha_0)}$$

$$= \frac{\alpha_0 r_c}{(R_{\text{IN}}+1/j\omega c)\left[1+\dfrac{r_c(1-\alpha_0)(R_{\text{IN}}+R_L+1/j\omega c)}{R_L(R_{\text{IN}}+1/j\omega c)}\right]}$$

$$= \frac{\alpha_0 r_c}{R_{\text{IN}}+1/j\omega c+\dfrac{r_c(1-\alpha_0)}{R_L}(R_{\text{IN}}+R_L+1/j\omega c)}$$

$$= \frac{\alpha_0}{\left[\left(1-\alpha_0\right)\left(1+\dfrac{R_{\text{IN}}}{R_L}\right)+\dfrac{R_{\text{IN}}}{r_c}\right]+\dfrac{1}{j\omega c}\left[\dfrac{1}{r_c}+\dfrac{(1-\alpha_0)}{R_L}\right]}.$$

Hence as ω is reduced the reactive term becomes important and reduces the ratio of output to input current. The effect of the emitter-decoupling capacitance has been ignored but it is usually much less important than the coupling capacitance.

The high frequency response may be obtained from the hybrid π equivalent circuit derived in Fig. 4.16 and shown in Fig. 6.18. This may be reduced to the circuit shown in Fig. 6.19.

Applying Kirchhoff's laws

$$v_{b'e} = (g_m v_{b'e} - i)Z_2 - iZ_3$$

$$v_{b'e}(g_m Z_2 - 1) = i(Z_2+Z_3).$$

In practice $g_m Z_2 \gg 1$.

Therefore

$$v_{b'e} = i\left(\frac{Z_2+Z_3}{g_m Z_2}\right)$$

or

$$i = \frac{v_{b'e} g_m Z_2}{Z_2+Z_3}.$$

Also

$$i_{IN} = \frac{v_{b'e}}{Z_1} + g_m v_{b'e} - i = v_{b'e}\left(\frac{1}{Z_1} + g_m - \frac{g_m Z_2}{(Z_2+Z_3)}\right).$$

Therefore

$$\frac{i}{i_{IN}} = \frac{g_m Z_2/(Z_2+Z_3)}{1/Z_1 + g_m - g_m Z_2/(Z_2+Z_3)} = \frac{g_m}{1/Z_1 + Z_3/Z_1 Z_2 + g_m Z_3/Z_2}.$$

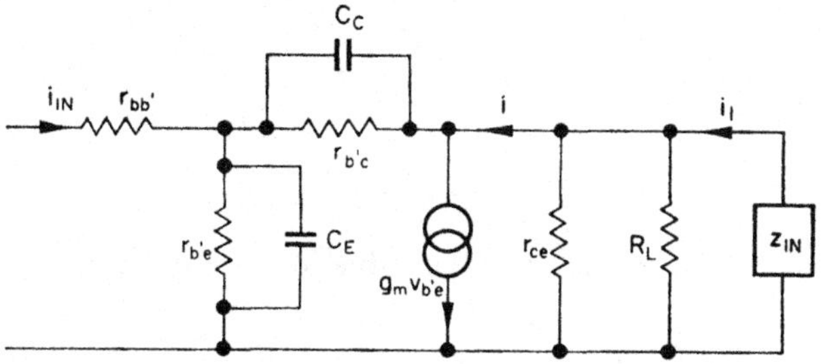

FIG. 6.18. H.F. equivalent circuit of Fig. 6.15.

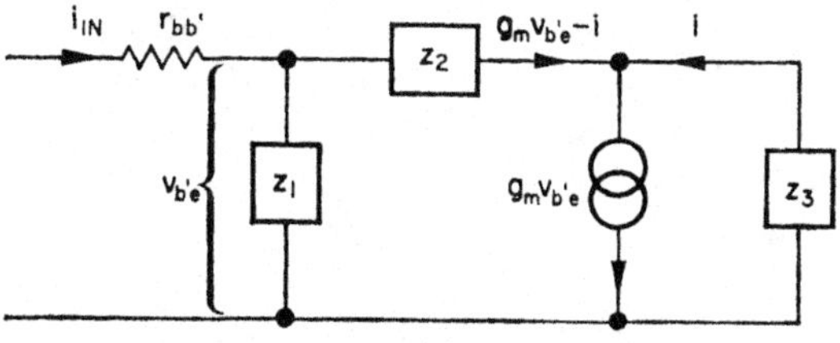

FIG. 6.19. Simplified H.F. equivalent circuit.

Hence the proportion of the current flowing through Z_{IN} may be evaluated and i_1/i_{IN} determined.

In practice Z_2 is very much greater than Z_1, and the fall off in current gain is due mainly to the reduction of Z_1 as the frequency increases. This reduction in impedance is due to the shunt capacitance c_e, the main part of which is due to the diffusion capacitance, and corresponds to the reduction in current gain derived in Chapter 4. Thus at high frequency

$$\alpha = \alpha_0/1+j\omega c_e r_e$$

$$\beta = \frac{\alpha_0/1+j\omega c_e r_e}{1-\alpha_0/1+j\omega c_e r_e}$$

$$= \frac{\alpha_0}{(1-\alpha_0)+j\omega c_e r_e}$$

$$= \frac{\alpha_0/1-\alpha_0}{1+j\omega c_e r_e/(1-\alpha_0)}$$

$$= \frac{\beta_0}{1+j\omega c_e r_e/(1-\alpha_0)}.$$

This may be compared to the reduction in Z_1 at high frequency, i.e.

$$\frac{r_{b'e}}{1+j\omega c_e r_{b'e}} = \frac{r_{b'e}}{1+[j\omega c_e r_e/(1-\alpha_0)]}, \quad \text{since } r_{b'e} = \frac{r_e}{1-\alpha_0}.$$

For improved response the diffusion capacitance must be reduced either by reducing the base width or increasing the rate of diffusion of charge carriers across the base. The former method, now becoming widely used, is constructed by epitaxial deposition as described in Chapter 3, while an example of the second method is the drift transistor.

6.10. Drift Transistor

In this type of transistor a base region is formed with an exponential impurity variation. This is produced by heating lightly doped *n* type germanium in an atmosphere with a strong arsenic impurity content. Arsenic diffuses into the germanium giving an impurity distribution, as shown in Fig. 6.20. Etching one side away gives an exponential distribution as shown in Fig. 6.21 and the emitter and collector junctions are alloyed in.

The " free " electrons in the base region diffuse through the base, resulting in a charge distribution as shown in Fig. 6.22.

The built-in field produced accelerates holes injected from the emitter to the collector, reducing the hole storage in the base and hence decreasing the diffusion capacitance.

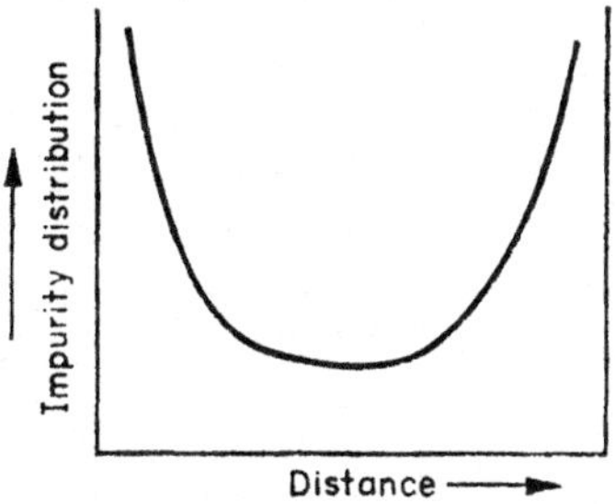

FIG. 6.20. Initial impurity distribution in drift transistor.

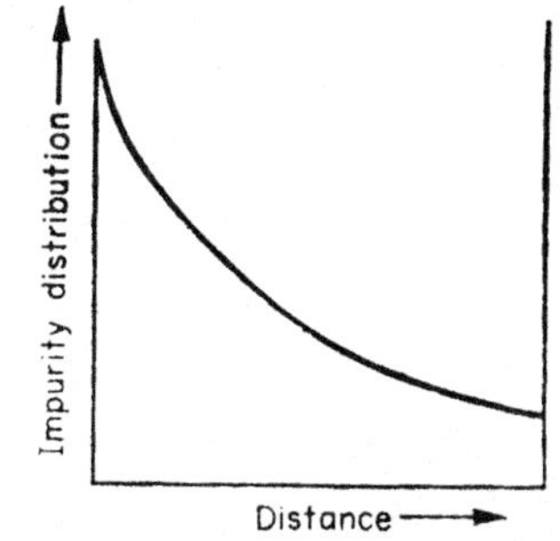

FIG. 6.21. Impurity distribution in drift transistor after etching.

6.11. R.F. Amplifiers

Radio-frequency amplifiers, operating over narrow frequency bands, usually have tuned collector loads and a typical circuit is shown in Fig. 6.23. The main considerations in the design of such amplifiers are the use of transistors with high cut-off frequencies, incorporation of stray and transistor capacities in tuned circuits, and effective neutralization.

High cut-off frequencies, particularly with grounded base circuits, may be obtained using epitaxial or drift transistors as mentioned previously, and amplifiers operating at a few hundred

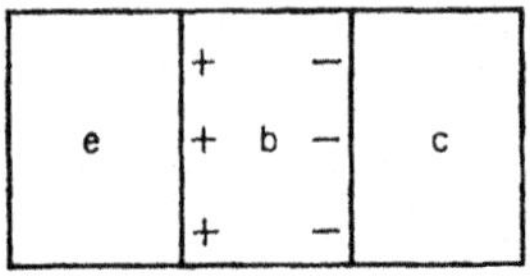

FIG. 6.22. Charge distribution in drift transistor.

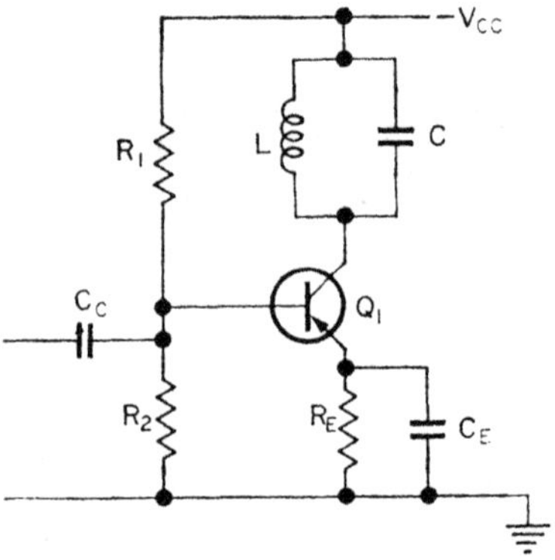

FIG. 6.23. R.F. voltage amplifier.

megahertz are fairly common. Output capacities may be incorporated in the tuned collector load and if necessary the input capacity may be included in the tuned input circuit. The use of a tuned input circuit is often advisable to obtain correct matching as shown in Fig. 6.24.

At high frequencies internal feedback in the transistor is troublesome. The energy fed back may cause instability or oscillation and any neutralizing network must counteract this.

Such a network is designed so that the voltage at the input due to the neutralizing network is of the same magnitude, but in anti-phase, to that produced by internal feedback.

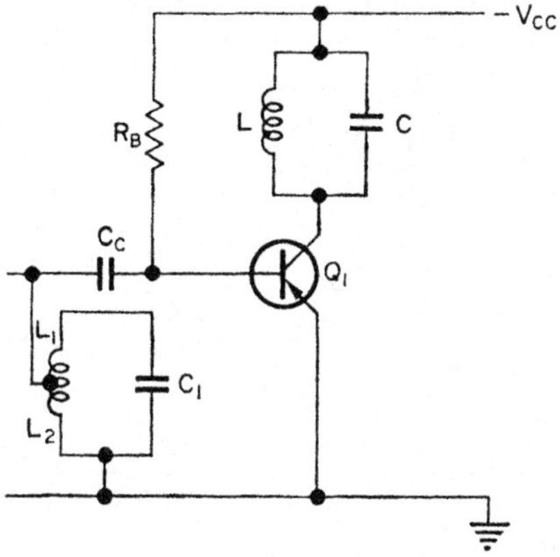

FIG. 6.24. R.F. voltage amplifier with tapped input circuit.

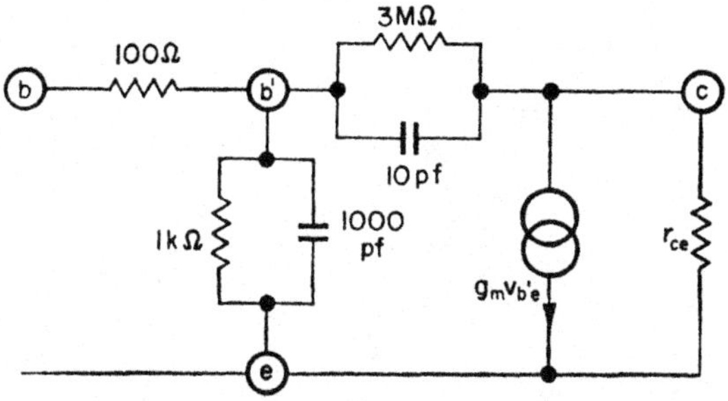

FIG. 6.25. Hybrid π equivalent circuit of R.F. voltage amplifier.

Appropriate neutralizing networks may be derived from the hybrid π equivalent circuit shown in Fig. 6.18. If we assume the typical values shown in Fig. 6.25, at 500 kHz

$$Y_{bb'} = 0 \cdot 01 \text{ siemen.}$$
$$Y_{b'e} = 0 \cdot 001 + j0 \cdot 00314 \text{ siemen.}$$
$$Y_{b'c} = 0 \cdot 33 \times 10^{-6} + j31 \cdot 4 \times 10^{-6} \text{ siemen.}$$

This may be converted into the normal π circuit shown in Fig. 6.26 by means of the T–π transformation and the path giving rise to feedback has an admittance

$$Y_{bc} = \frac{Y_{bb1}Y_{b'c}}{Y_{bb'}+Y_{b'e}+Y_{b'c}}$$

$$= 7{\cdot}87\times10^{-6}+j26{\cdot}4\times10^{-6}.$$

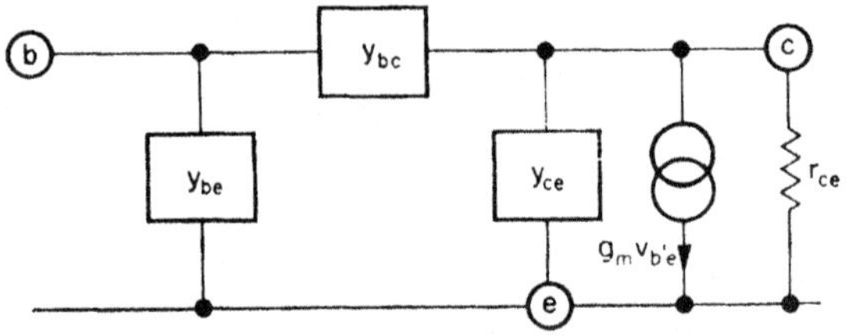

FIG. 6.26. Normal π equivalent circuit of R.F. voltage amplifier.

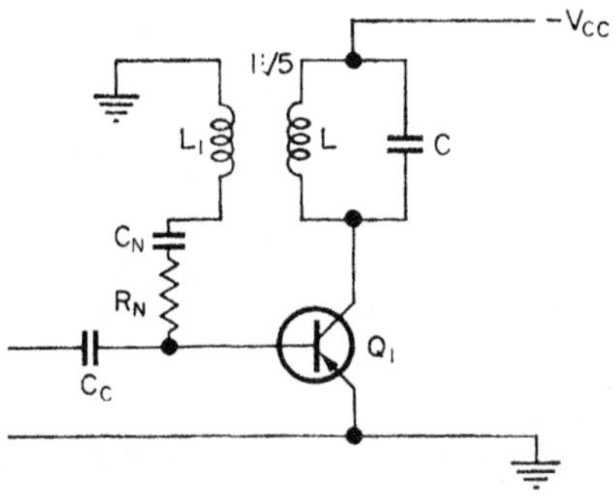

FIG. 6.27. Neutralized R.F. voltage amplifier.

At the particular frequency of 500 kHz, this may be replaced by a series network $10{,}400+j3500$, i.e. a resistance of 10·4 kΩ in series with a capacitor of 9·1 pF.

The neutralized circuit is shown in Fig. 6.27, the coupling being assumed to give a step-down of $\sqrt{5}:1$ and 180° phase shift. The values of C_N and R_N are now 45·5 pF and 2·08 kΩ and the energy fed back via this network exactly balances that fed back internally through the transistor.

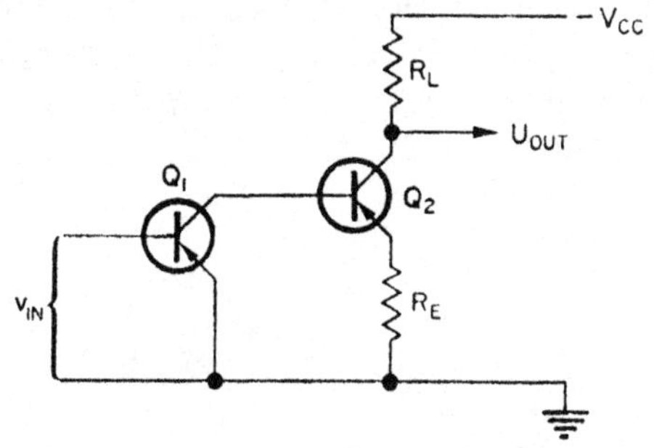

FIG. 6.28. Simple d.c. amplifier.

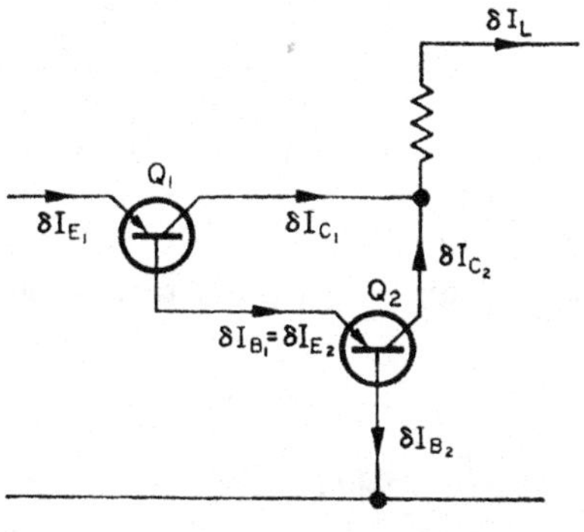

FIG. 6.29. D.C. amplifier using common base connection.

6.12. D.C. Amplifiers

The low frequencies response of amplifiers is limited by the size of coupling and decoupling capacitors and eventually direct coupling has to be used. Directly coupled amplifiers have the inherent disadvantage of drift and it is often necessary to use suitable compensating circuits.

The simplest form of directly coupled amplifier is shown in Fig. 6.28 and comprises two common emitter stages. The emitter resistance in the second stage enables correct bias to be obtained and also provides feedback. Although negative feedback reduces the drift, it reduces the gain by the same amount and does not therefore compensate for drift.

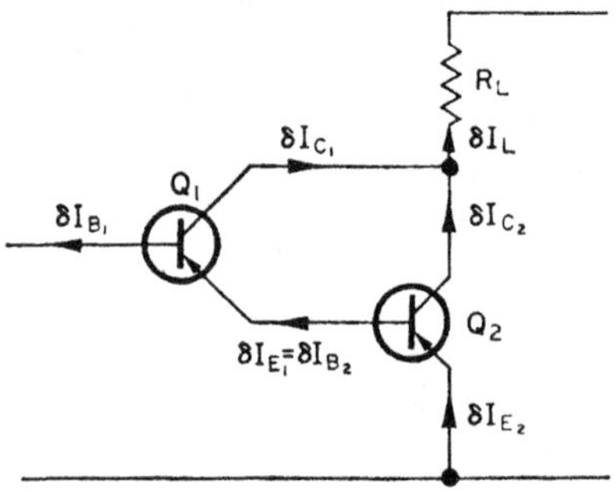

FIG. 6.30. D.C. amplifier using common emitter connection.

An alternative circuit is shown in Fig. 6.29 and this enables an effective current gain of almost unity to be obtained. From the circuit

$$\begin{aligned}\delta I_L &= \delta I_{c_1} + \delta I_{c_2}\\ &= \alpha_1 \delta I_{E_1} + \alpha_2 \delta I_{E_2}\\ &= \alpha_1 \delta I_{E_1} + \alpha_2 \delta I_{B_1}\\ &= \alpha_1 \delta I_{E_1} + \alpha_2 \delta I_{E_2}(1-\alpha_1).\end{aligned}$$

Effective current gain

$$\frac{\delta I_L}{\delta I_{E_1}} = \alpha_1 + \alpha_2 - \alpha_1\alpha_2$$

If $\alpha_1 = \alpha_2 = 0{\cdot}98$, $\delta I_L/\delta I_{E_1} = 1$ (very nearly).

The grounded emitter circuit corresponding to Fig. 6.29 is shown in Fig. 6.30.

Then

$$\begin{aligned}\delta I_L &= \delta I_{c_1} + \delta I_{c_2} \\ &= \beta_1 \delta I_{B_1} + \beta_2 \delta I_{B_2} \\ &= \beta_1 \delta I_{B_1} + \beta_2 \delta I_{E_1} \\ &= \beta_1 \delta I_{B_1} + \beta_2 (1+\beta_1) \delta I_{B_1}.\end{aligned}$$

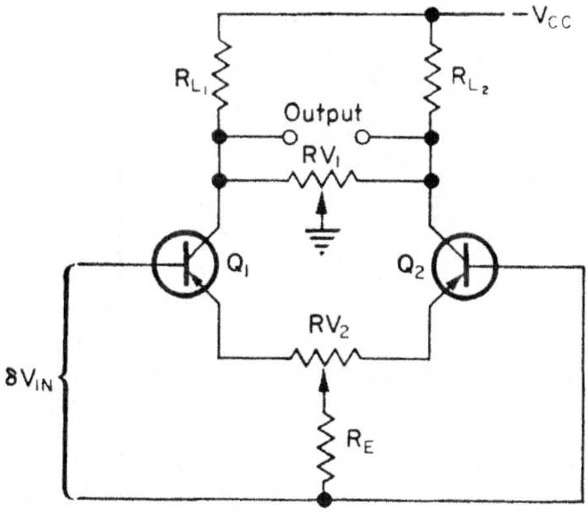

FIG. 6.31. Compensated d.c. amplifier.

Effective current gain

$$\frac{\delta I_L}{\delta I_{B_1}} = \beta_1 + \beta_2 + \beta_1 \beta_2.$$

If $\alpha_1 = \alpha_2 = 0{\cdot}98$, $\beta_1 = \beta_2 = 49$ and $\delta I_L / \delta I_{B_1} = 2500$.

The circuits shown on Figs. 6.29 and 6.30 may be extended to include more transistors and further variations may be obtained by using a combination of *pnp* and *npn* types.

Drift may be corrected using a circuit similar to that shown in Fig. 6.31. In this circuit the drift due to one transistor is balanced by that produced in the other and a balanced output is obtained. The two potential dividers control the division of emitter current and the voltage gains of the two transistors, and by successive adjustment a circuit may be obtained that is relatively independent of temperature.

Where very small changes are involved the problem of drift becomes more important and recourse has to be made to a chopper type amplifier. In this type direct voltages and currents are

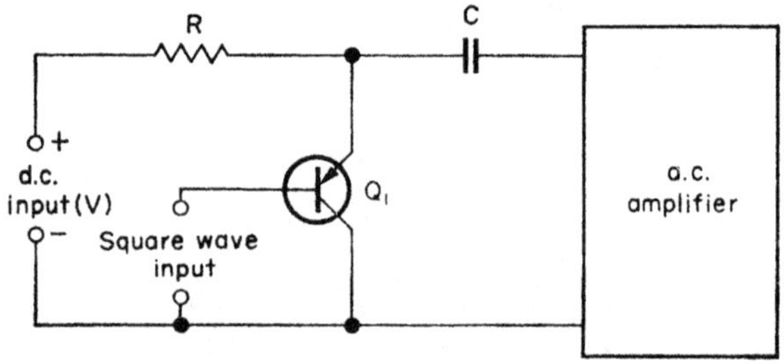

FIG. 6.32. Chopper type d.c. amplifier.

interrupted at regular intervals and a square wave whose amplitude is proportional to the direct input is produced. This output wave may then be amplified by conventional amplifiers.

The simplest form of chopper circuit is shown in Fig. 6.32. The d.c. input voltage V to be measured is connected through a resistance R to the emitter of a transistor and the input of a conventional amplifier. The base of the transistor is fed with a square wave sufficient to drive the transistor from the full-on to the cut-off condition. Thus the voltage at the emitter varies between the limits of approximately zero and V. This is amplified by the conventional amplifier and an output obtained whose magnitude is proportional to the input voltage. If a d.c. output is required a suitable R.C. integrating network may be connected to the output of the amplifier, but this must be switched in and out of circuit in synchronism with the square wave on the base of the transistor.

Questions for Chapter 6

Q.6.1. Compare and contrast the three methods of connection used in transistor circuits.

Q.6.2. Compare methods of attaining the correct operating point in valve and transistor circuits. Mention briefly the importance of bias stabilization in transistor circuits. Determine the d.c. operating point for the circuit shown in Fig. 6.33. Comment on the circuit and note all assumptions made.

$I_{co} = 10\ \mu A$, $\alpha_0 = 0{\cdot}98$, $R_L = 3{\cdot}3\ k\Omega$, $R_B = 220\ k\Omega$, $R_E = 470\ \Omega$.

(I.E.E. Applied Electronics, Part III, 1963)

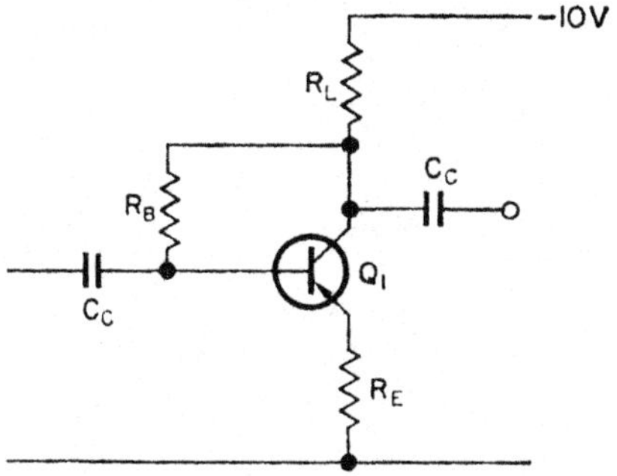

FIG. 6.33. Circuit for Q.6.2.

Q.6.3. Explain why grounded emitter amplifiers require some form of d.c. stabilization whereas grounded base amplifiers are usually unstabilized.

Describe a method of stabilizing the collector current and derive an expression for the factor of stability of the circuit chosen.

Q.6.4. Measurements made on a small *pnp* transistor gave the following results when connected as a common emitter circuit.

V_{CE}	−2 V					−8 V				
I_c mA	0·1	1	2	3	4	0·3	1·4	2·6	3·9	5·9
I_B μA	0	20	40	60	80	0	20	40	60	80

Construct load lines for 2000 Ω and 500 Ω loads and a 6 V supply. Determine suitable operating points for an input of 40 μA peak to peak in each case. Determine also the corresponding current and voltage outputs.

Q.6.5. The transistor whose characteristics are given in question 6.4 is used in the circuit shown in Fig. 6.34. Determine the output voltage if the alternating base input is 40 μA peak to peak and the standing bias is 20 μA.

Determine also the overall current gain of the stage if it is connected to another stage of input resistance 500 Ω by a capacitor of very low reactance.

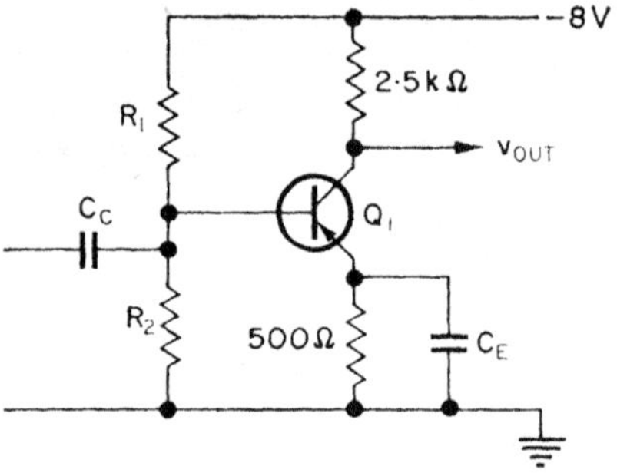

FIG. 6.34. Circuit for Q.6.5.

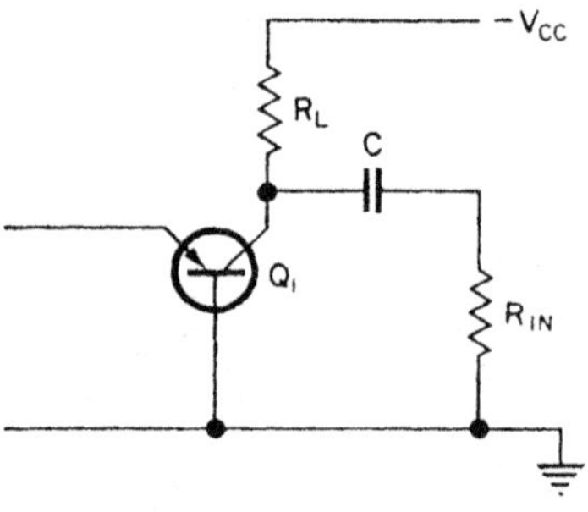

FIG. 6.35. Circuit for Q.6.8.

Q.6.6. Calculate after deriving the necessary formula the input resistance of a common collector transistor stage with emitter load resistance of 5 kΩ. The working point is such that $r_e = 40$ Ω, $r_b = 500$ Ω, $r_c = 1$ MΩ, $\alpha_0 = 0{\cdot}98$. Show an arrangement for obtaining the necessary base bias current that will not result in an excessive lowering of the input impedance.

(I.E.E. Electronic Engineering, Part III, 1958)

Q.6.7. Prove the relations shown in Table 6.1 stating any assumptions made.

Q.6.8. Prove that for the common base amplifier shown in Fig. 6.35, the output current falls by 3 db when

$$\omega = \frac{1}{C\left[R_{\mathrm{IN}} + \left(\frac{r_c R_L}{r_c + R_L}\right)\right]}.$$

Assume a constant input current and a collector resistance much greater than the base resistance.

CHAPTER 7

Power Amplifiers

7.1. Classification of Power Amplifiers

It is convenient to classify power amplifiers either in terms of the drive conditions, or in terms of the load. From the point of view of the input current, the amplifiers may be classified as follows (assuming amplifiers in common emitter mode):

Class A—Collector current flows during the whole of the input cycle.

Class B—Collector current flows during half the input cycle.

Class C—Collector current flows for less than half of the input cycle.

We may refer to the " angle of flow " of an amplifier, where this angle is proportional to the fraction of one input current cycle for which output current flows from one transistor. So that, for instance, a class C amplifier has an angle of flow which is less than 180°. Or a class AB amplifier which has a standing bias intermediate between class A and class B, when used with large input signals will have an angle of flow rather more than 180°.

From the load point of view it is clear that in an amplifier having only one transistor, with a resistive load, any operation other than class A will produce distortion in the output waveform from a sinusoidal input. So distortionless class AB and class B amplification is only possible, with a non-tuned load, if two transistors are used in a " push-pull " arrangement.

On the other hand, if a tuned load is used, with a relatively high Q factor, this will select only the narrow range of frequencies, near the resonant frequency to which the circuit is tuned, and will therefore largely eliminate distortion. So the class C tuned amplifier, with a single transistor, is a possibility for radio frequency work.

Of these possibilities we shall consider the class B audio frequency amplifier in most detail, as this is a very common application of transistors. These amplifiers may be considered either as voice frequency amplifiers, for the approximate frequency range from 20 Hz to 20 kHz, or as servo-amplifiers driving the control winding of an electrical machine, and usually operating at a single, fixed frequency (which is often 400 Hz).

TABLE 7.1

I_B (mA)	I_C (mA) for $V_{ce} = -0{\cdot}5$ V	I_C (mA) for $V_{ce} = -5$ V
0·2	9	12
0·6	28	35
1·0	48	57
1·4	66	77

7.2. Class A Power Amplifier

In this class of amplifier the transistor is biased such that it conducts over the whole of the input cycle swing. The maximum power is dissipated within the transistor when there is no input signal. With transformer coupling the maximum output power is half the maximum dissipation, so that the maximum theoretical efficiency, with an ideal transistor and loss-free transformer, is 50%. With a resistive load directly in the collector lead the maximum efficiency of a class A amplifier is 25%.

The use of a load line with transistor characteristics was considered in Chapter 6. We will here consider a numerical example to show how the values of the output and dissipation powers may be calculated.

A small *pnp* transistor is used in the circuit of Fig. 7.1. The collector characteristics for this transistor are given in Table 7.1 and drawn on the graph of Fig. 7.2. The source resistance is $R_S = 100\ \Omega$, the load resistance is $R_L = 50\ \Omega$ and the biasing

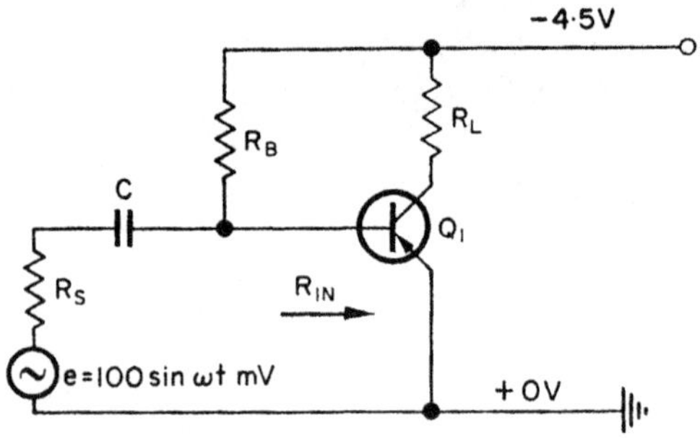

FIG. 7.1. Transformer-coupled class A amplifier.

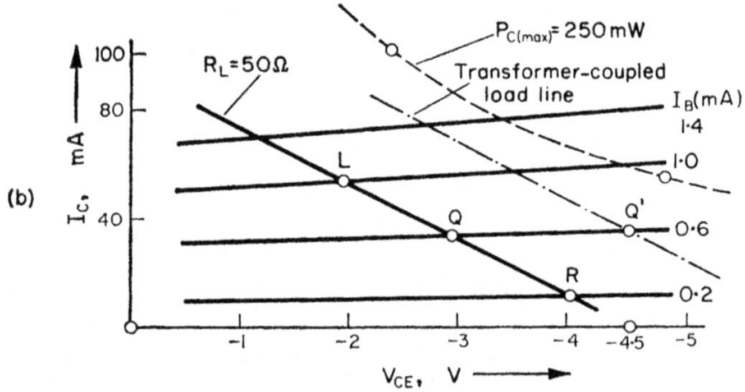

FIG. 7.2. Collector characteristics with load line.

resistor is $R_B = 7{\cdot}2\ \text{k}\Omega$. The available sinusoidal signal is $e = 100 \sin \omega t$ millivolts. The reactance of the coupling capacitance C is negligible at the working frequency of the amplifier. The base–emitter d.c. potential is set at 0·18 V and the a.c. input resistance, which may be estimated from the input characteristic, is typically about 150 Ω. Knowing the supply voltage is −4·5 V then the standing base current is

$$I_B = \frac{4{\cdot}5-0{\cdot}18}{7{\cdot}2}\ \text{mA} = 600\ \mu\text{A}.$$

The load line corresponding to 50 Ω is drawn on the graph of Fig. 7.2. As there is no emitter resistor in this particular circuit, the whole supply voltage is available across the transistor (at the instant of zero current) and the load line is drawn from the −4·5 point on the voltage axis. The operating point, about which we may consider the swing of a.c. quantities, is given by the intersection of the load line and the characteristic corresponding to $I_B = 0{\cdot}6$ mA. This point is labelled Q on the graph.

The peak input voltage at the base may be calculated, ignoring the shunting effect of the bias resistor R_B, and is then

$$\hat{v}_{\text{IN}} = \frac{100 \times R_{\text{IN}}}{R_S + R_{\text{IN}}} = 100\left(\frac{1{\cdot}5}{2{\cdot}5}\right)\ \text{mV}.$$

Hence the peak base current signal is

$$\hat{\imath}_{\text{IN}} = \frac{\hat{v}_{\text{IN}}}{R_{\text{IN}}} = 100\left(\frac{1{\cdot}5}{2{\cdot}5}\right)\left(\frac{1}{150}\right)\ \text{mA}$$

or $$\hat{\imath}_{\text{IN}} = 400\ \mu\text{A}$$

giving $$i_{\text{IN(r.m.s.)}} = \frac{400}{\sqrt{2}}\ \mu\text{A} = 280\ \mu\text{A}.$$

Therefore the base current swings between 1 mA and 0·2 mA. The points where these characteristics intersect the load line are indicated as L and R on the graph. The peak to peak values of collector current and voltage are found by projection, horizontally and vertically, from these points L and R. The graph shows that the peak to zero values are

$$\hat{\imath}_c = \frac{52-12}{2} = 20\ \text{mA}.$$

$$\hat{v}_c = \frac{3{\cdot}9-2{\cdot}1}{2} = 0{\cdot}9\ \text{V}.$$

Or the r.m.s. output values are

$$i_{c(\text{r.m.s.})} = \frac{20}{\sqrt{2}} = 14 \text{ mA}$$

and

$$v_{c(\text{r.m.s.})} = \frac{0\cdot 9}{\sqrt{2}} = 0\cdot 63 \text{ V}.$$

Hence the output power in the 50 Ω load is

$$P_O = 14 \times 0\cdot 63 = 9 \text{ mW}.$$

With no signal input the dissipation within the transistor is

$$P_C = I_C V_{CE} = 32 \times 3 = 96 \text{ mW},$$

which corresponds to the total power available at the collector. Under the conditions used in this calculation the efficiency of the stage, ignoring bias current, is

$$\eta_{\text{coll}} = \frac{\text{a.c. output}}{\text{available collector power}} = \frac{9}{96} 100\% = 9\cdot 4\%.$$

Also the overall efficiency is

$$\eta_{\text{total}} = \frac{\text{a.c. output}}{\text{d.c. input}} = \frac{9 \times 100\%}{32 \times 4\cdot 5} = 6\cdot 2\%$$

which takes account of the standing dissipation in the resistive load.

The input power to the base is

$$P_{\text{IN}} = [i_{\text{IN(r.m.s.)}}]^2 R_{\text{IN}} = (0\cdot 28)^2 (150)\ \mu\text{W} = 12\ \mu\text{W}.$$

Therefore the power gain of the stage is

$$G_P = \frac{9 \times 10^3}{12} = 750 \text{ times}$$

or, in decibels,

$$G_P = 10 \log_{10} \left(\frac{P_o}{P_{\text{IN}}} \right) = 28\cdot 7 \text{ db}.$$

Greater output power may be achieved in two ways. Firstly, increased driving current at the base will give a greater output swing. This is limited, by the knee voltage of the transistor and the supply voltage, to a voltage swing of 4 V peak to peak in this case.

Secondly, transformer coupling will leave practically the whole supply voltage available for the collector to emitter voltage. If we now assume a load of 10 Ω coupled into the collector through a transformer with a 2·2:1 step-down ratio, this gives an effective a.c. load resistance of 50 Ω. Ignoring any d.c. potential across the transformer, the primary operating point becomes Q', corresponding to $V_{CE} = -4{\cdot}5$ V and $I_B = 0{\cdot}6$ mA. The load line passing through this point Q' will be parallel to the original 50 Ω line. If the base drive is not increased, then the collector efficiency will be almost equal to the overall efficiency without the transformer. With no signal input the collector dissipation corresponding to point Q' is $P_C = 4{\cdot}5 \times 34 = 150$ mW.

If the standing base current is increased to 1 mA to allow for a larger input signal, then the no-signal dissipation will rise to $(4{\cdot}5 \times 57)$ mW $= 250$ mW, which may well be the maximum permissible dissipation for a small germanium transistor. Also the peak collector current produced by the peak of the input current will be greater than 100 mA, which may be excessive.

Assuming that the base drive is increased to give a base current peak swing of 0·6 mA on either side of the operating point Q', then the available alternating collector current and voltage will be

$$i'_{c(\text{r.m.s.})} = \frac{60-34}{\sqrt{2}} = 18{\cdot}4 \text{ mA}$$

and

$$v'_{c(\text{r.m.s.})} = \frac{4{\cdot}5-3{\cdot}2}{\sqrt{2}} = 0{\cdot}92 \text{ V}.$$

Then the output power is $P_O = (18{\cdot}4)0{\cdot}92$ mW $= 17$ mW and the overall efficiency is

$$\eta = \frac{17}{150}\,100\% = 11{\cdot}3\%.$$

From these calculations it becomes clear that the maximum output and efficiency, with least distortion, is obtained from a load line which passes through the knee voltage point, at maximum permissible collector current, with the operating point just below the maximum collector dissipation curve. The curve corresponding to $P_{c(max)} = 250$ mW is plotted as a dashed line on the characteristics of Fig. 7.2. Thermal stability, on the lines indicated in the previous chapter, is a major consideration in such amplifiers.

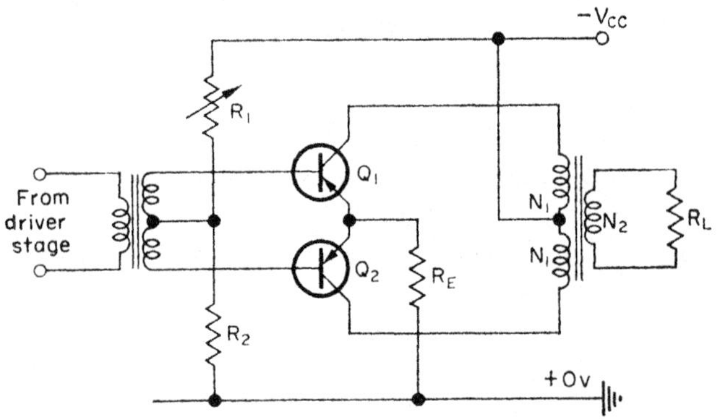

FIG. 7.3. Transformer-coupled push-pull amplifier.

7.3. Common Emitter Push-pull Amplifier

A pair of matched transistors may be used as a " push-pull " combination, with an angle of flow in each transistor at any point between 360° (class A) and 180° (class B). The basic circuit for a transformer-coupled push-pull amplifier, using two similar transistors in the common emitter mode, is given in Fig. 7.3. It is usual to have the base resistor R_1 adjusted to give a small standing base current, so that both transistors are just conducting with no input signal. This means that the amplifier is operating in the near class B (or class AB) state and the power dissipation with no input signal, under quiescent conditions, will be small. The output transformer provides a transformation of the load resistance to give maximum power transfer for minimum distortion.

The class A push-pull amplifier, which includes the class AB amplifier with suitably small signals has the advantage of giving power amplification with very low distortion. These advantages, compared with the possibility of using two transistors in series or in parallel, may be summarized as follows:

(i) For iron-cored transformers there is no resultant direct magnetization, as the direct current in each half of the primary of the output transformer flows in opposite directions. So no special precautions are required to prevent magnetic saturation of the output transformer.

(ii) Any alternating current flowing in the supply line from another source (such as a power supply with poor smoothing) will not affect the output, since the output voltage is proportional to the instantaneous difference between the two collector currents. Any disturbances on the common line affect each transistor equally.

(iii) With a centre-tapped transformer in the output, under ideal conditions, even harmonics of the signal frequency are eliminated.

To achieve these advantages the two transistors must be accurately matched in all characteristics, and the input and output transformers must be accurately centre-tapped. Matching of the transistor pair involves several factors. The current gains can be generally matched on the peak value to within 10%. Differences in the slope of the curve relating current gain and emitter current results in a difference between the mean current gains, leading to distortion. Also, the two transistors should respond in an identical manner to changes of signal frequency and ambient temperature. The cut-off frequencies must be matched in order to balance the gain of the higher order signal frequencies. Any difference in the extrinsic base resistance (which is independent of emitter current) would result in differences in the input current to the two halves of the amplifier. However, this last effect is generally swamped by the generator resistance.

The effective elimination of the second harmonics may be understood if we consider the centre-tap of the input transformer as at earth potential. Then the two transistor bases are supplied in antiphase from a balanced voltage source. At the instant when one base is at positive peak potential then the other base will be at negative peak potential.

If the voltage at the base of Q_1 is $\hat{v}\sin\omega t$, then we will assume that the collector current, due to curvature of the input characteristic of the transistor, is of the form

$$i_{c1} = a_1+b_1 \sin \omega t+c_1 \sin 2\omega t+d_1 \sin 3\omega t+\ldots .$$

The corresponding voltage at the base of Q_2 will be $\hat{v}\sin(\omega t+180°)$ and the collector current will be

$$i_{c2} = a_2+b_2 \sin(\omega t+180°)+c_2 \sin 2(\omega t+180°) +d_2 \sin 3(\omega t+180°)+\ldots$$

or

$$i_{c2} = a_2-b_2 \sin \omega t+c_2 \sin 2\omega t-d_2 \sin 3\omega t+\ldots .$$

The resulting output voltage is proportional to the difference current in the output transformer, and therefore,

$$v_{\text{OUT}} \propto (i_{c1}-i_{c2}).$$

If the constants of the two transistors are equal (with $a_1 = a_2$, etc.), then

$$v_{\text{OUT}} \propto 2\times(b \sin \omega t+d \sin 3\omega t+\ldots).$$

However, if the transistor constants are not equal then even harmonics will be present in the output.

At the same time current flowing in the supply line is the sum of the two collector currents, giving

$$i_s = i_{c1}+i_{c2} = 2(a+c \sin 2\omega t+\ldots).$$

This shows that, under ideal conditions, there is no current in the supply line at the fundamental signal frequency, and this is also an important feature of the class A push-pull amplifier and the associated power supply requirements.

The class B push-pull amplifier does not enjoy the above-mentioned advantages, but does make high efficiency operation possible with a resistive load. Class B operation has a theoretical efficiency of 78% and an actual efficiency of 70% is possible. In addition the power drain in no-signal conditions is insignificant.

To calculate the effective load resistance of a particular power amplifier we need to consider the output transformer in more detail, using the sketch of Fig. 7.4. In that figure N_1 is the number of turns **in each half** of the primary winding (so that the

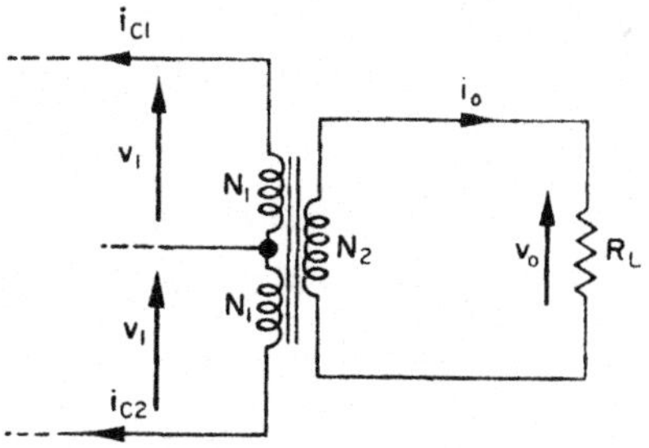

FIG. 7.4. Output transformer and turns ratio.

total number of primary turns is $2N_1$). The number of secondary turns, on the load side of the transformer, is N_2. The currents to the collectors of the transistors, through separate halves of the primary winding, are i_{c1} and i_{c2}. In the class B amplifier either i_{c1} or i_{c2} may be taken as zero at any instant. The resultant primary ampere-turns are $N_1\ (i_{c1}-i_{c2})$. If the output voltage and current are v_o and i_o respectively, then the following equations are true for a loss-free transformer:

$$\frac{2v_1}{2N_1} = \frac{v_1}{N_1} = \frac{v_o}{N_2} \tag{1}$$

and

$$N_1(i_{c1}-i_{c2}) = N_2 i_o. \tag{2}$$

Dividing eqn. (1) by eqn. (2),

$$\frac{v_1}{N_1^2(i_{c1}-i_{c2})} = \frac{v_o}{N_2^2 i_o}$$

or

$$\frac{v_1}{i_{c1}-i_{c2}} = \left(\frac{v_o}{i_o}\right)\left(\frac{N_1}{N_2}\right)^2 = R_L\left(\frac{N_1}{N_2}\right)^2. \quad (3)$$

Hence the load line may be of a slope which is the reciprocal of $R_L(N_1/N_2)^2$ drawn on a characteristic of instantaneous difference current plotted against collector voltage v_1. The effective resistance $R_L(N_1/N_2)^2$ is known as the " composite dynamic resistance ".

Multiplying together the two equations (1) and (2) we have that,

$$v_1(i_{c1}-i_{c2}) = v_o i_o.$$

Using r.m.s. values of voltage and current, then

$$P_O = v_1(i_{c1}-i_{c2}) \text{ watts.} \quad (4)$$

Equation (4) shows that the output power is given by the product of the r.m.s. voltage across one half of the transformer primary and the r.m.s. difference current.

7.4. Design Example—Transformer-coupled Push-pull Amplifier

As a first step we may ignore the " knee voltage " and assume a power transistor with ideal linear characteristics. Then the maximum peak output voltage, on the primary side of the transformer, is equal to the supply voltage V_{cc}. If the collector to emitter load per transistor is R_{CE} ohms, then the maximum peak current is V_{CC}/R_{CE} amperes.

Each transistor of the push-pull pair supplies an exact half-cycle of the output waveform, when biased to exactly the class B condition. This means that a transistor is conducting on alternate half-cycles. So the maximum output power from the class B push-pull amplifier is

$$P_{o(\max)} = \left[\frac{V_{cc}}{\sqrt{2}}\right]^2 \left(\frac{1}{R_{CE}}\right) = \frac{V_{cc}^2}{2R_{CE}} \text{ watts.}$$

The input power to each transistor is the product of the supply voltage and the mean current, where

$$i_{\text{mean}} = \left(\frac{V_{cc}}{R_{CE}}\right)\left(\frac{1}{\pi}\right) \quad \text{(as shown in Section 5.2).}$$

Therefore the total input power is

$$P_{\text{IN}} = \frac{2}{\pi}\left(\frac{V_{cc}}{R_{CE}}\right) V_{cc} = \frac{2V_{cc}^2}{\pi R_{CE}} \text{ watts.}$$

This enables us to calculate the maximum theoretical efficiency as

$$\eta_{\max} = \frac{P_{o(\max)}}{P_{\text{IN}}} = \frac{V_{cc}^2}{2R_{CE}} \Big/ \frac{2V_{cc}^2}{\pi R_{CE}}$$

or

$$\eta_{\max} = \frac{\pi}{4} = 78{\cdot}5\%.$$

This equation shows that, in approximate terms, we may consider the dissipation within each transistor, at maximum output, as $\frac{1}{10} P_{o(\max)}$. Assuming a sinusoidal input signal, a graph relating the amplitude of the input signal, the collector dissipation, and the efficiency of an ideal class B amplifier, is given as Fig. 7.5. Note that this graph shows that the maximum dissipation, within each transistor, occurs when the peak collector current is 0·64 of the maximum allowable value and the power efficiency is 50%.

For calculation we shall consider a class B push-pull amplifier with a supply voltage (V_{cc}) of -24 V and transistors with a maximum permissible collector current of 4 A. The graph of Fig. 7·6 shows that the suitable load line for each transistor is

$R_{CE} = 24/4 = 6\ \Omega$. Therefore approximate values for the powers involved are

$$P_{o(\max)} = \frac{(24)^2}{2 \times 6} = 48\ \text{W}.$$

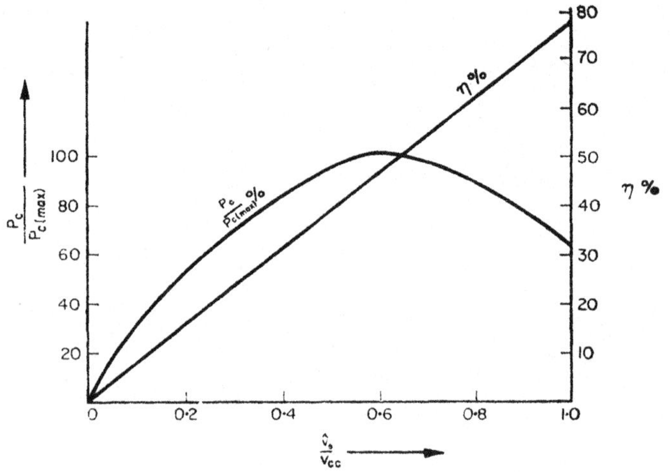

FIG. 7.5. Graph relating efficiency, power dissipation and signal amplitude for class B amplifier.

and the power drawn from the d.c. supply,

$$P_{\text{IN}} = \frac{2(24)^2}{\pi \times 6} \simeq 60\ \text{W},$$

$$\text{or dissipation per transistor} \simeq \frac{60-48}{2} = 6\ \text{W}.$$

In fact, due to the collector saturation resistance, each transistor has a knee voltage, which is dependent on the collector current. To avoid non-linearity, which would arise if the voltage were allowed to swing into the knee voltage region of the collector characteristics, the minimum collector to emitter voltage must be maintained slightly higher than the knee voltage. This region is defined by the dashed line, at an angle ψ to the vertical, and the vertical current axis, in Fig. 7.6. On a current/voltage graph $\tan\psi$ will have the dimensions of a resistance. For this example $\tan\psi = 0{\cdot}4\ \Omega$ is a suitable value. So the peak output voltage is reduced to

$$\hat{v}_o = 24\left(\frac{R_{CE}}{R_{CE}+\tan\psi}\right) = \left(\frac{6}{6{\cdot}4}\right)24 = 22{\cdot}5\ \text{V}.$$

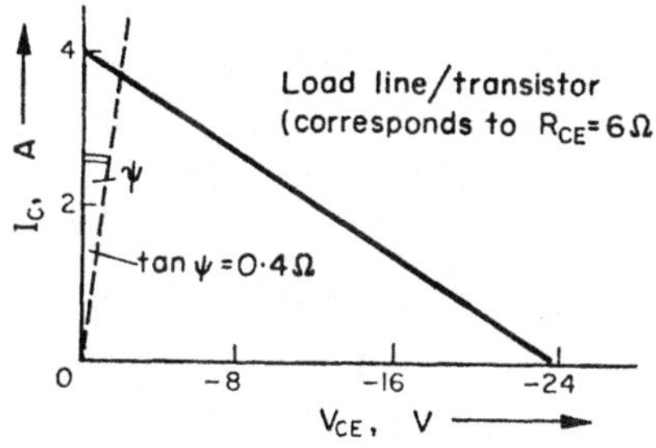

FIG. 7.6. Load line and effect of knee voltage.

and the peak output current becomes

$$\hat{\imath}_o = \frac{V_{cc}}{R_{CE}+\tan\psi} = \frac{24}{6{\cdot}4} = 3{\cdot}75\ \text{A}.$$

This gives an available output power of

$$P_{o(\max)} = \frac{22{\cdot}5\times 3{\cdot}75}{2} = 42\ \text{W}.$$

Additionally, there is an emitter resistor in the collector to emitter loop and this forms an effective load which is not bypassed, so $R_{CE} = R_E+R_C$. If $R_E = 0{\cdot}4\ \Omega$, then the useful reflected load is

$$R_C = 6-0{\cdot}4 = 5{\cdot}6\ \Omega.$$

The useful output power is $43(5{\cdot}6/6{\cdot}0) = 39$ W and the dissipation in R_E is 3 W.

To avoid crossover distortion which occurs due to the curvature of the transistor input characteristics at low values of base current, a small quiescent bias current flows in each transistor at all times. This current might be $I_Q = 100$ mA in the example chosen. The voltage across each transistor will vary with the signal conditions, but the quiescent current will increase transistor dissipation by a factor of at least $\frac{1}{5} I_Q V_{cc}$, to give a total additional power dissipation of

$$P_Q = 0{\cdot}04 \times 24 = 1 \text{ W}.$$

Total dissipation in the two transistors is therefore

$$P_{\text{diss}} = 60 - 42 + 1 = 19 \text{ W}.$$

The overall efficiency of the push-pull amplifier may be seen to be

$$\eta = \frac{\text{a.c. output power}}{\text{d.c. input power}} = \frac{39}{61} = 64\%$$

with a power dissipation of nearly 10 W in each transistor and 3 W in the common emitter resistor.

The drive current required at the base of each transistor to produce the maximum output power is simply the peak collector current divided by the minimum expected current gain of the transistors. In this case the approximate drive current required is

$$\hat{\imath}_{\text{IN}} = \frac{3{\cdot}75}{30} = 125 \text{ mA (peak)}.$$

7.5. Transformer-less Audio Amplifiers

The transformer is often the most bulky and expensive item in power amplifiers operating at audio frequencies. There is considerable incentive to design amplifiers, suitable for the efficient push-pull configuration, without transformers. The " Phase-splitter " circuit arrangement is often used as a driving stage before one or more push-pull power stages, and in this way the input transformer is eliminated. Figure 7.7 shows a phase-splitting amplifier driving one push-pull amplifier stage, where *npn* transistors are employed (a corresponding circuit using *pnp* transistors could equally well be used). The phase-splitter arrangement is not peculiar to transistors, but the circuit has a very simple

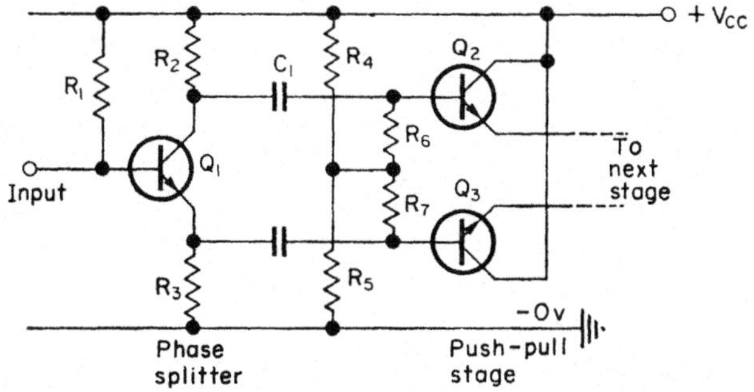

FIG. 7.7. Phase-splitter circuit.

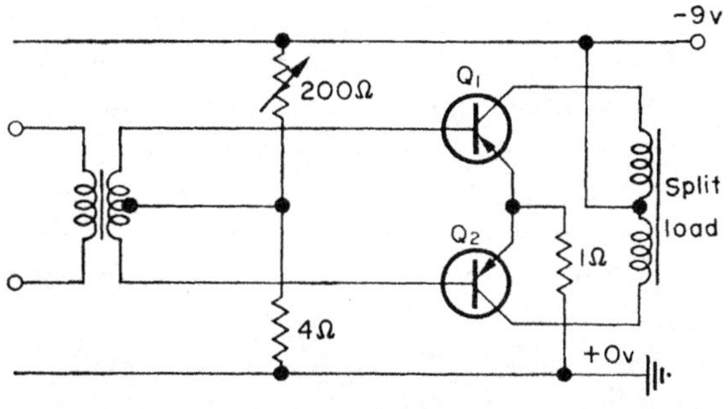

FIG. 7.8. Power amplifier with split-load winding.

symmetry in this case. The resistors R_2 and R_3 may be identical, and if Q_1 is a high gain transistor, the two voltage outputs will be equal in magnitude and opposite in phase, over the audio frequency range. The phase splitter is essentially a voltage amplifier and requires a load of fairly high impedance to give a reasonably true performance. The voltage gain is slightly less than one.

In some applications the output transformer of a push-pull power stage may be omitted by simply driving the output current into a centre-tapped load winding. This is possible where the output transformer is not required to give a step-down ratio and where the resulting load impedance matches that of the output transistors. As the output impedance of a power transistor in the common emitter mode is only a few ohms this is an attractive possibility. The circuit of Fig. 7.8 shows two germanium *pnp* power transistors working into a 10 Ω load. This circuit would give about 6 W, with the load either being a centre-tapped loud-speaker coil or the excitation winding of a servo motor. The r.m.s. drive power required is of the order of 300 mW and the d.c. resistance in each half of the secondary winding of the driving transformer should be about 4 Ω, for correct base bias.

A complementary pair of transistors (one *pnp* and one *npn*) can be used to great advantage in transformer-less power amplifiers. The complementary amplifier can thus drive a resistive load without incorporating any transformers but retain the high efficiency of near class B push-pull operation. There are several ways in which a pair of complementary transistors may be included in a push-pull circuit. Due to the difficulty of manufacturing matched high-power complementary transistors together with the relatively high gain and low cost of germanium *pnp* power transistors, it is usual to put the complementary pair in the driving stage and not in a high power output stage. These circuits are discussed in more detail with much useful design information by Tobey and Dinsdale (see Appendix B). For perfect operation the current/voltage curves of the *pnp* and *npn* transistors must have exact symmetry, over the whole working range of frequency and temperature. The circuit of a complete transformer-less power amplifier of a type which would give an output of 20 W with an efficiency of about 70%, is given in Fig. 7.9. In this circuit transistors Q_1 and Q_2 are medium power *pnp* transistors, while Q_3 is an *npn* transistor matched to Q_2, Q_4 and Q_5 are high power *pnp* transistors. The thermistor TH_1 provides a temperature sensitive bias to the base of Q_3. The resistors R_F

and R_B provide a.c. and d.c. feedback. When the bases of Q_2 and Q_3 are driven positive from Q_1, then Q_3 conducts and Q_2 is driven OFF. Conversely, in the negative-going half-cycle, transistor Q_2 is driven ON and Q_3 is non-conducting. Since the driving stage looks like a pair of emitter followers the voltage gain is slightly less than one and the current gain is that of the transistors involved.

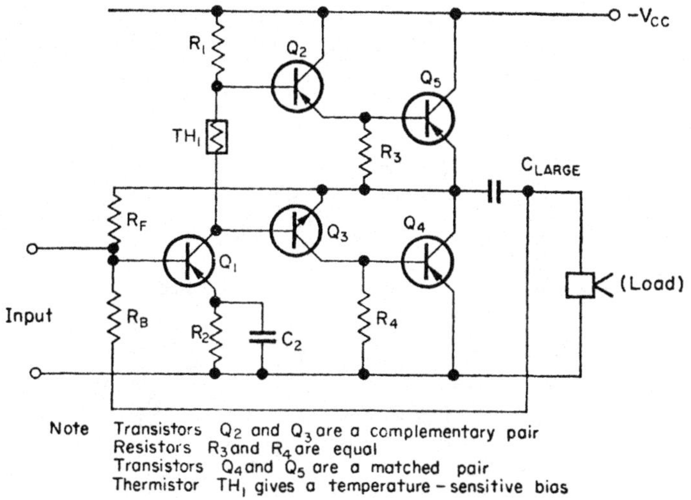

FIG. 7.9. Transformer-less power amplifier, with complementary transistors in the driving stage.

7.6. Servo System Amplifiers

Many circuits are available in which power transistors operate as switches. This leads to low power dissipation, with high power output, and good thermal stability at high ambient temperatures. A power amplifier can operate in the switched mode if the signal to be amplified and the power supply are at the same frequency. This is very suitable for servo systems, where the operating frequency is often 400 Hz, and the essentials of such an amplifier are given in Fig. 7.10. The power supply for the transistors is from a biphase bridge rectifier, without filtering. The supply voltage must be slightly greater than the peak signal voltage, by an amount corresponding to the knee voltage of the transistor. The contrast between the supply voltage in the conventional class B amplifier, with a smoothed supply voltage, and that in the class B switched servo amplifier is illustrated in the simple graphs (a)

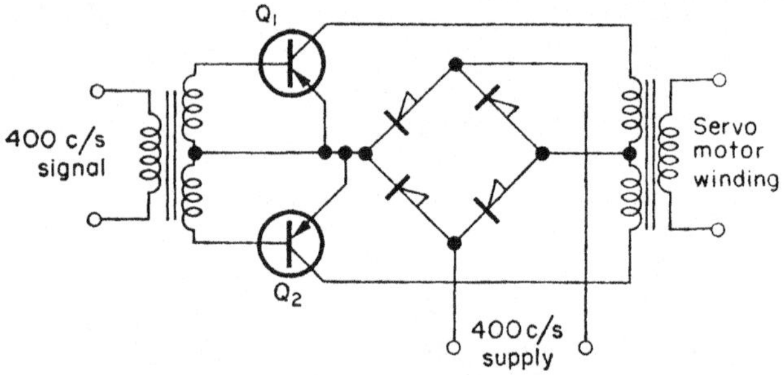

FIG. 7.10. Servo power amplifier with switched operation.

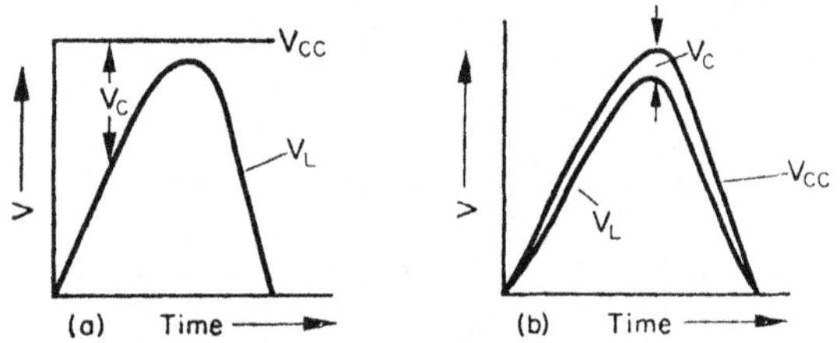

FIG. 7.11. Comparison of conventional and switched mode class B amplifier operation.

and (b) of Fig. 7.11. Clearly the dissipation in the transistors can be made quite small in the servo amplifier, merely allowing for the internal transistor volt drop V_C.

The resulting efficiencies of the two types of amplifier are compared in the graph of Fig. 7.12. This graph assumes a small knee voltage for the power transistors. An efficiency of 95% has been claimed for transistor amplifiers using the switched mode of operation at a single frequency. This suggests that an output of

200 W at 400 Hz is available with only 5 W dissipation in each transistor. For this purpose high gain germanium power transistors, which are readily available, would be quite suitable. Further details of the design of this class of amplifier are given in the reference quoted in Appendix B.

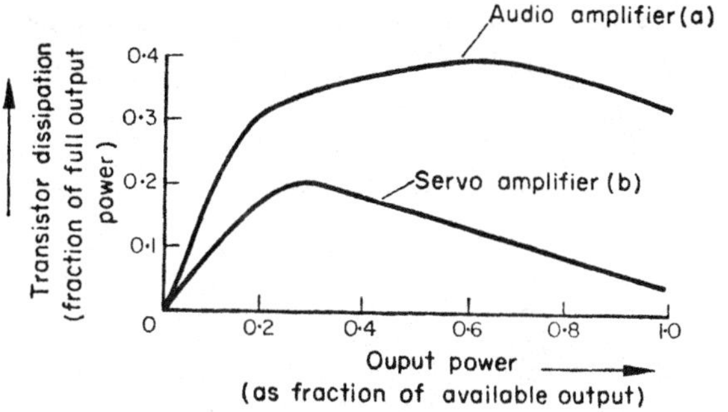

FIG. 7.12. Power dissipation in conventional and switched mode amplifiers.

7.7. Class C High Frequency Amplifiers

Tuned amplifiers operating in the class C mode are employed, particularly in radio transmitters, for large output powers, up to hundreds of kilowatts. To give a reasonable amplitude ratio of fundamental to harmonic frequencies, the tuned circuit load must have a Q factor of at least 7 and preferably of 10. The Q factor and dynamic resistance (at resonance) of a parallel tuned circuit are related by the equation $R_D = Q\omega_0 L$, where ω_0 is the angular resonant frequency (radians/second). Hence the available output power is

$$P_o = \frac{v_o^2}{R_D} \text{ watts}$$

where v_o is the r.m.s. output voltage. The r.m.s. voltage is related to the peak to peak voltage ($2\hat{v}_o$) by

$$v_o = \frac{\hat{v}_o}{\sqrt{2}}.$$

Here the peak to peak voltage is somewhat less than twice the supply voltage (V_{cc}).

But the maximum permissible collector to emitter voltage is strictly limited for any transistor and is often less than 50 V for a germanium power transistor. These considerations limit the available output power from a class C amplifier using a power transistor, and in practice these amplifiers are very useful for small transmitters with output powers of a few watts. The power

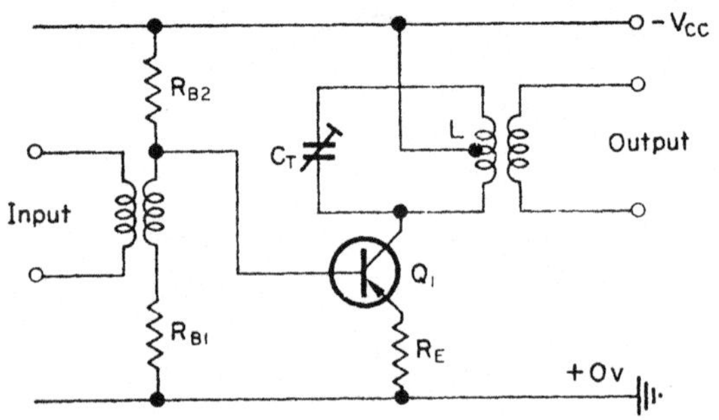

FIG. 7.13. Simple class C amplifier.

efficiency of the class C amplifier is rather difficult to calculate theoretically and will not be considered in detail here. One method is given by Parker (see Appendix B).

A calculation based on the simple class C amplifier of Fig. 7.13 will illustrate some aspects of the design problems. An available germanium power transistor can withstand a maximum V_{CE} of 60 V, at a maximum collector current of 6 A, at which the knee voltage will be less than 1 V. The maximum power dissipation, when the transistor is bolted down to a 6 in. square aluminium heat sink, is 6 W if the ambient temperature does not exceed 50°C.

The operating frequency of the amplifier is 500 kHz and the loaded Q factor of the circuit is 8, with a total tuning capacitance of 250 pF. We can choose the supply voltage to be $V_{cc} = -54$ V and assume a maximum peak to zero voltage swing of 50 V. The dynamic resistance of the tuned circuit load is given by the expression

$$R_D = \frac{Q}{\omega_0 C} \text{ ohms}$$

where $\omega_0 = 2\pi f_o = 10^6\pi$ rad/sec.

Therefore $$R_D = \frac{8}{(10^6\pi)(250 \, . \, 10^{-12})} \, \Omega \simeq 10 \text{ k}\Omega.$$

The maximum available r.m.s. voltage is

$$v_o = \frac{50}{\sqrt{2}} = 35 \text{ V}.$$

Hence the available output power is $(35)^2/10^4$ W or

$$P_{o(max)} = \frac{1230}{10^4} \simeq 0{\cdot}12 \text{ W}.$$

Clearly, a worth-while power output may only be obtained by increasing the voltage swing (which is impossible with this germanium transistor) or by decreasing the dynamic resistance. R_D is dependent on the capacitance to inductance ratio, and it may be possible, within limits, to increase the tuning capacitance and thus have less inductance in the circuit, and yet retain a reasonable Q factor. In this example the tuning capacitance might possibly be increased to 2000 pF, which corresponds to an associated inductance of 50 μH, at a frequency of 500 kHz. The eightfold increase of tuning capacitance would give a similar reduction in the dynamic resistance and the available output power would be almost 1 W. It would seem difficult to achieve a greater output power than 1 W at 500 kHz when the supply voltage must be limited to less than 60 V.

Questions for Chapter 7

Q.7.1. State the meaning of class A, B, and C operation as applied to power amplifiers. What is meant by angle of flow?

A class C transistor amplifier operates at 10 MHz, with a collector load tuning capacitance of 1000 pF and a resultant Q factor of 12·6. The d.c. supply is 40 V, the collector dissipation is 1 W and the collector efficiency is 75%.

Calculate approximate values for the following:

(a) effective dynamic resistance of the tuned load circuit;
(b) power fed to the load;
(c) direct current drawn from the supply; and
(d) the peak-to-peak value of the output voltage. (H.N.D.)

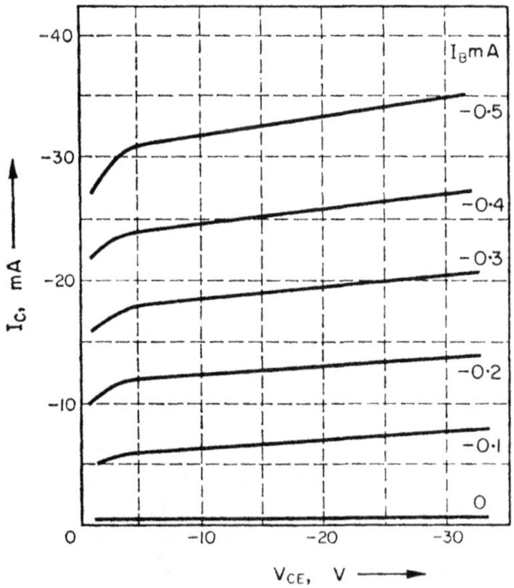

FIG. 7.14. Transistor characteristics for Q.7.3.

Q.7.2. Define the following terms with reference to a tuned-collector R.F. amplifier, class C operation, collector efficiency, collector dissipation, angle of flow.

Such an amplifier is tuned to $4 \times 10^7/2\pi$ Hz. The supply voltage is 36 V, the collector dissipation is 0·5 W and the collector efficiency is 80%. The tuning capacitor is 500 pF and the operating Q factor is 5.

Calculate the a.c. power fed to the load, the mean current taken by the stage, and the peak-to-peak value of the output voltage. (H.N.C.)

Q.7.3. A *pnp* germanium transistor is used in the common emitter connection with an ideal 3:1 step-down transformer in the collector circuit. The operating point is determined by a resistor connected between the base and the negative supply.

If the supply is 16 V and the collector dissipation is not to exceed 250 mW, determine graphically a suitable operating point to give approximately maximum " undistorted " output in a load resistor of 75 Ω connected across the secondary of the transformer. Calculate the power in the load resistor and the value of the bias resistor. The transistor characteristics given in Fig. 7.14 should be used for this question. (I.E.E., Nov. 1962)

Q.7.4. Discuss the advantages of a push-pull audio frequency amplifier using a pair of complementary transistors. Also point out the problems associated with using *pnp* and *npn* transistors together in this way.

A push-pull common emitter amplifier stage has complementary transistors, working into a resistive load, with linear characteristics and correct d.c. bias conditions. However, the current gains of the two transistors are 50 and 40 respectively. Show how the output waveform may be derived from the d.c. characteristics of the transistors and estimate the percentage harmonic distortion if this may be assumed to be only second harmonic. (H.N.C.)

Q.7.5. Draw the circuit for a push-pull amplifier stage using two *pnp* transistors. Assume an ideal transformer with a resistive load and transistor characteristics given by

$$i_{c1} = 300\,(v_{be1}-0{\cdot}1)+10(v_{be1}-0{\cdot}1)^2$$

and

$$i_{c2} = 250\,(v_{be2}-0{\cdot}11)+9(v_{be2}-0{\cdot}11)^2$$

where collector current is in mA and base to emitter voltage in V. Determine the percentage second harmonic distortion at the output. The base bias voltage is 220 mV at each transistor and the peak sinusoidal signal voltage is 100 mV to each base. Comment on the significance of your results.

Q.7.6. Sketch circuit diagrams to show two methods which use a pair of transistors in a power output stage without the use of an output transformer. Explain the principles of operation, and compare the properties of the two circuits.

CHAPTER 8

Oscillator Circuits

THERE are many transistor oscillator circuits in use but in the space of one chapter it would be impossible to attempt complete coverage. In this chapter examples of each type are discussed and analysed. Inverter and converter circuits are also considered as a form of oscillator.

8.1. L.C. Feedback Oscillator

In this type of oscillator damped oscillations initiated in a tuned circuit are maintained and increased, to a limiting value by positive feedback. A proportion of the output power is fed back to the input of the transistor, which converts sufficient d.c. power into a.c. power to overcome losses in the circuit and load. Figures 8.1, 8.2, 8.3 and 8.4 show examples of this type of oscillator but only the first and fourth will be considered in detail.

8.1.1. *The tuned collector oscillator*

In the circuit shown as Fig. 8.1, an oscillation in the tuned circuit will induce a current in the base circuit and an amplified version of this current will appear at the collector. The transformer and transistor each produce 180° phase shift and the current flowing from the collector is in the correct phase to maintain oscillations.

An approximate analysis of the tuned collector oscillator may be made by assuming sinusoidal operation and the simplified equivalent T circuit shown in Fig. 8.5. In this circuit the frequency of oscillation is assumed low enough to ignore high frequency

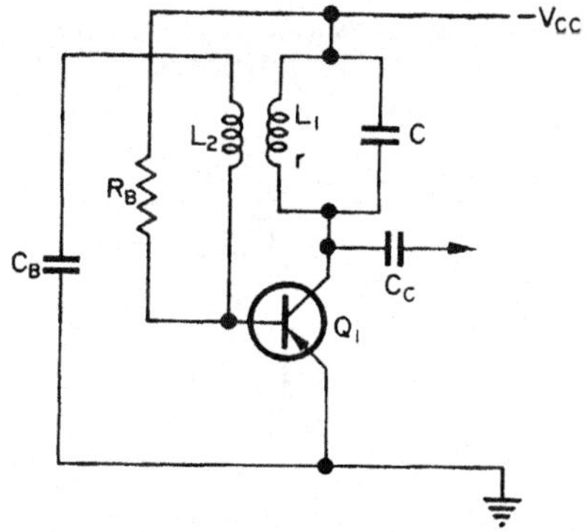

FIG. 8.1. Tuned collector oscillator.

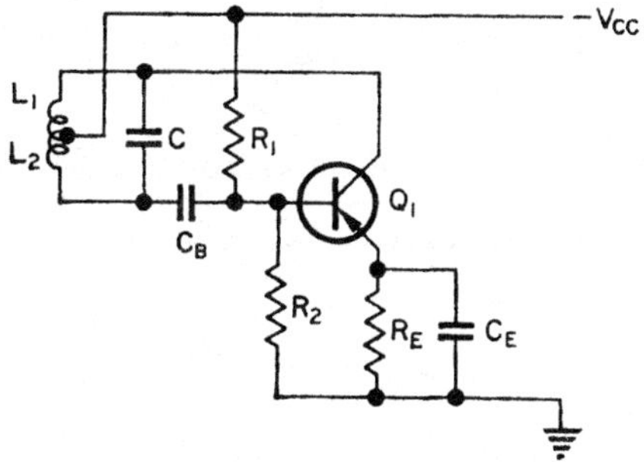

FIG. 8.2. Hartley transistor oscillator.

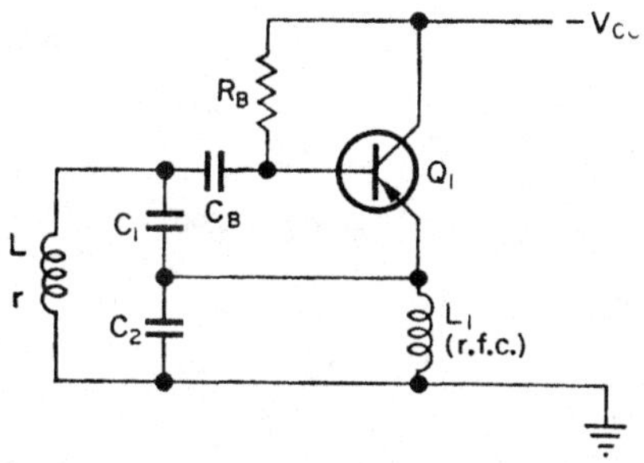

FIG. 8.3. Colpitts transistor oscillator.

effects within the transistor and the collector resistance is assumed high enough to be neglected. The components R_B and C_B are only included for d.c. biasing purposes and have been omitted from the equivalent circuit.

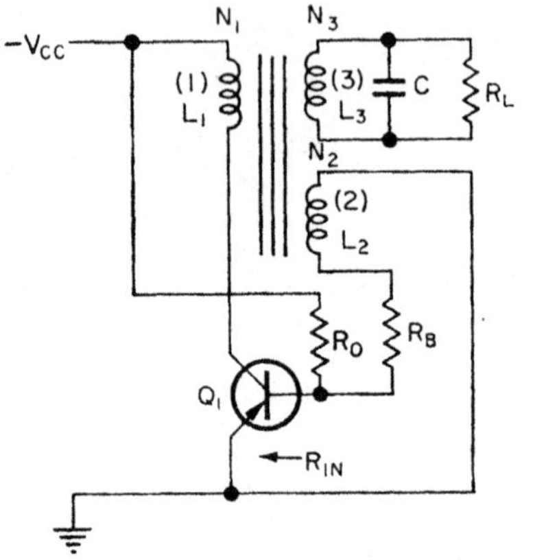

FIG. 8.4. Oscillator with inductively coupled tuned circuit.

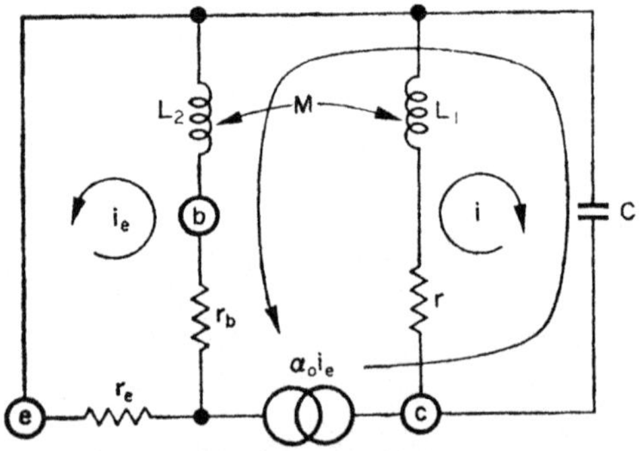

FIG. 8.5. Equivalent circuit of tuned collector oscillator.

Applying Kirchhoff's laws to the equivalent circuit,

$$i_e(r_e+r_b+j\omega L_2)-\alpha_0 i_e(r_b+j\omega L_2)\pm j\omega Mi = 0,$$

$$i\left(r+j\omega L_1+\frac{1}{j\omega C}\right)-\alpha_0 i_e\cdot\frac{1}{j\omega C}\pm j\omega Mi_e(1-\alpha_0) = 0.$$

Putting $R = r_e + r_b(1-\alpha_0)$,

$$i_e\{R+j\omega L_2(1-\alpha_0)\} = \mp j\omega M i,$$

$$i_e\left\{\frac{\alpha_0}{j\omega C} \mp j\omega M(1-\alpha_0)\right\} = \left(r+j\omega L_1+\frac{1}{j\omega C}\right) i.$$

Hence

$$\left\{R+j\omega L_2(1-\alpha_0)\right\}\left\{r+j\omega L_1+\frac{1}{j\omega C}\right\}$$

$$= \mp j\omega M\left\{\frac{\alpha_0}{j\omega C} \mp j\omega M(1-\alpha_0)\right\}.$$

Equating real parts,

$$Rr-\omega^2 L_1 L_2(1-\alpha_0)+\frac{L_2}{C}(1-\alpha_0) = \mp\frac{\alpha_0 M}{C}-\omega^2 M^2(1-\alpha_0).$$

If $M = \sqrt{L_1 L_2}$,

$$Rr+\frac{L_2}{C}(1-\alpha_0) = \mp\frac{\alpha_0 M}{C}.$$

This equation gives the condition of oscillation and obviously the positive value must be taken,

i.e. $$M = \frac{CRr}{\alpha_0}+L_2\left(\frac{1-\alpha_0}{\alpha_0}\right).$$

The imaginary parts lead to

$$R\omega L_1-\frac{R}{\omega C}+\omega L_2 r(1-\alpha_0) = 0$$

or $$\omega^2 = \frac{1}{L_1 C\left\{1+\frac{L_2}{L_1}\cdot\frac{r}{R}(1-\alpha_0)\right\}}.$$

This expression gives the frequency of oscillation and indicates that the term

$$\frac{L_2}{L_1}\cdot\frac{r}{R}(1-\alpha_0)$$

should be as low as possible for best stability.

The above analysis has ignored the effects of load impedance on the oscillator and the factors that control the amplitude of the oscillation. The derived equations only apply to the initial build up since the oscillations are limited by non-linearity of the characteristics. In practice transistor oscillators usually operate under class AB, class B or class C conditions. If the loading is small, its effects may be simulated by an increase in the value of the coil resistance r; the operation under class AB conditions is illustrated in the next section.

8.1.2. *Oscillator with inductively coupled tuned circuit*

The circuit of an oscillator of this type is shown in Fig. 8.4. Due to the tuned circuit the voltages across each winding of the coupling transformer will be approximately sinusoidal even though the current waveform for the base circuit will be truncated.

If we assume the voltage across winding (2) is $V_2 \cos \omega t$ and current flows over the whole cycle, the alternating base current is given by

$$i_b = \frac{R_o}{R_{\text{IN}}+R_o} \cdot \frac{V_2 \cos \omega t}{R_B+R_{\text{IN}}R_o/(R_{\text{IN}}+R_o)} = \frac{R_o V_2 \cos \omega t}{R_B(R_{\text{IN}}+R_o)+R_{\text{IN}}R_o},$$

where R_o is the base bias resistor, R_{IN} is the input resistance of the transistor, and R_B is the series resistance in the feedback winding,

or
$$i_b = \frac{V_2 \cos \omega t}{R_T},$$

where

$$R_T = R_B + R_{\text{IN}} + \frac{R_B R_{\text{IN}}}{R_o}.$$

If the angle of flow is 2θ, the waveform is as shown in Fig. 8.6 and Fourier Analysis gives a fundamental component

$$i_{b_1} = \frac{K V_2 \cos \omega t}{R_T}, \quad \text{where } K = \frac{1}{2\pi}(\theta - \sin 2\theta).$$

If β_0 is the current gain of the transistor, the fundamental component of the collector current is given by

$$i_{c_1} = \beta_0 i_{b_1}.$$

The voltage across winding (1) is then

$$V_1 = Ri_{c_1},$$

where R is the total resistance referred to winding (1),

i.e. $$V_1 = R\beta_0 i_{b_1} = \frac{R}{R_T} \cdot \beta_0 K V_2 \cos \omega t.$$

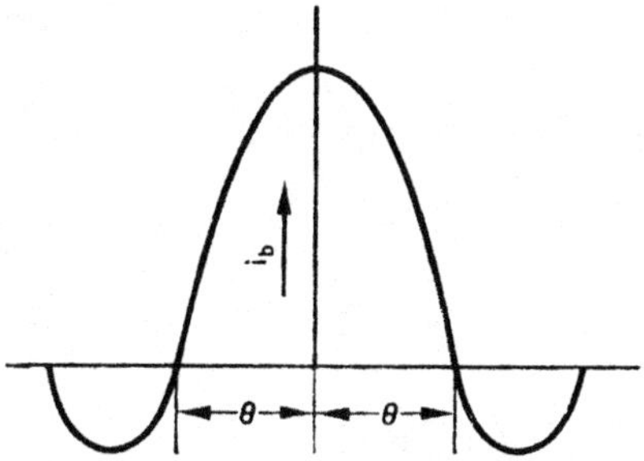

FIG. 8.6. Base current waveform for inductively coupled oscillator.

Hence for sustained oscillations

$$V_2 = \frac{n_2}{n_1} V_1,$$

where n_2/n_1 is the turns ratio,

i.e. $$V_2 \cos \omega t = \frac{n_2}{n_1} \cdot \frac{R}{R_T} \cdot \beta_0 K V_2 \cos \omega t$$

or $$\beta_0 = \frac{n_1 R_T}{n_2 R K}.$$

This expression gives the condition for oscillations to be maintained at a frequency close to the resonant frequency of the tuned circuit, i.e. $1/2\pi\sqrt{(L_3C)}$.

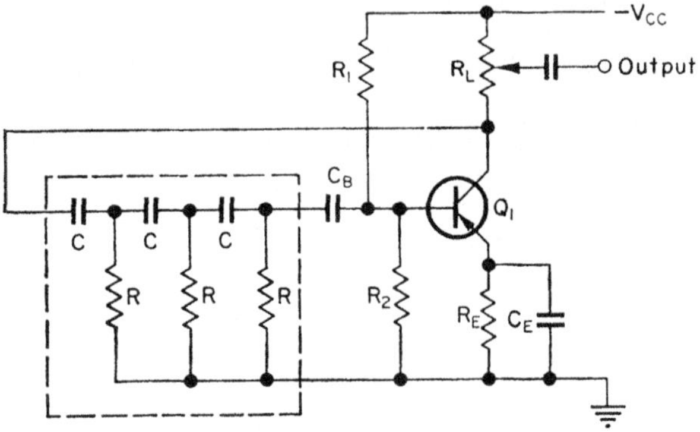

FIG. 8.7. Single-stage R.C. oscillator with voltage feedback.

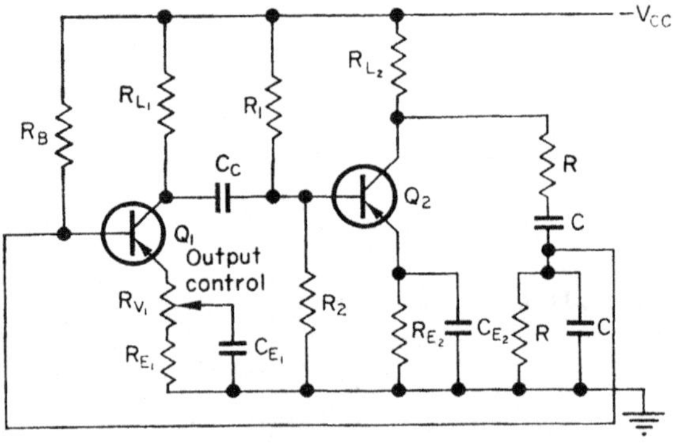

FIG. 8.8. Two-stage R.C. oscillator with voltage feedback.

8.2. R.C. Oscillators

These oscillators rely for their operation on the phase shift produced by resistance–capacitance networks and no tuned circuit is involved. Since the frequency of operation is inversely proportional to the capacity (not the square root of the capacity as in the case of L.C. oscillators), wide ranges may be obtained for a given capacity variation. Figures 8.7, 8.8, 8.9 and 8.10 are typical

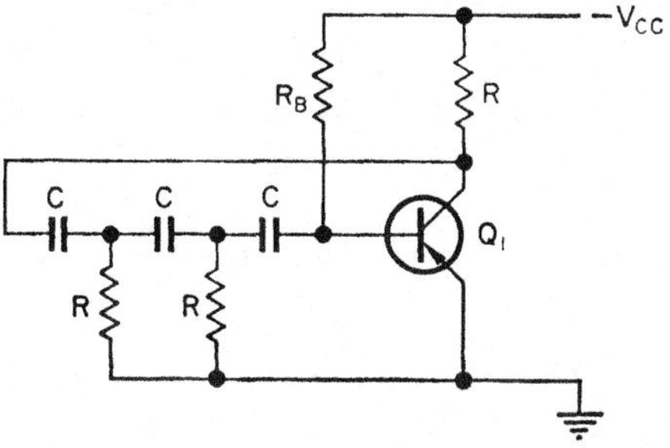

FIG. 8.9. Single-stage R.C. oscillator with current feedback.

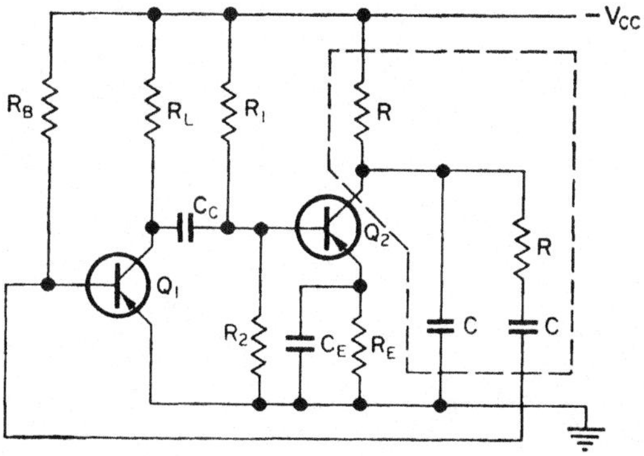

FIG. 8.10. Two-stage R.C. oscillator with current feedback.

examples of this type, the first two relying on voltage feedback and the second two on current feedback. It has been the authors' experience that the latter type are easier to design and require less components.

8.2.1. *Single-stage R.C. oscillator with voltage feedback*

The circuit of this oscillator is shown in Fig. 8.7 and relies for its operation on that part of the circuit enclosed in dotted lines and reproduced in Fig. 8.11.

If i_1, i_2 and i_3 are the currents circulating in each loop, the following equations apply:

$$v_{IN} = i_1\left(R+\frac{1}{j\omega C}\right)-i_2R \tag{1}$$

$$0 = i_2\left(2R+\frac{1}{j\omega C}\right)-i_1R-i_3R \tag{2}$$

$$0 = i_3\left(2R+\frac{1}{j\omega C}\right)-i_2R \tag{3}$$

$$v_{OUT} = i_3R. \tag{4}$$

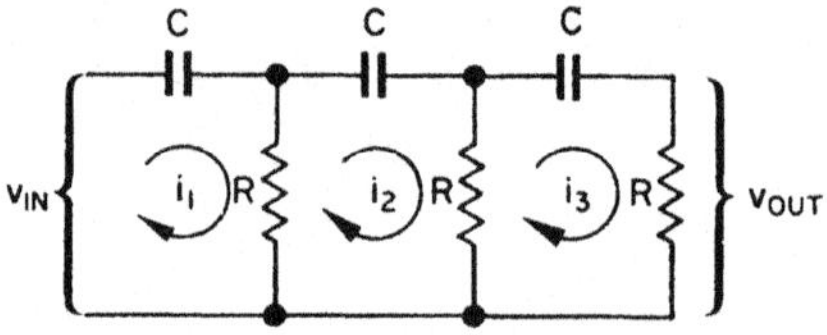

FIG. 8.11. Feedback network for single-stage R.C. oscillator.

Equation (3) gives

$$i_2 = i_3\left(2+\frac{1}{j\omega CR}\right).$$

Equation (2) gives

$$i_1 = i_3\left(2+\frac{1}{j\omega CR}\right)\left(2+\frac{1}{j\omega CR}\right)-i_3R$$

$$= i_3\left\{\left(3-\frac{1}{\omega^2C^2R^2}\right)+\frac{4}{j\omega CR}\right\}.$$

Equation (1) gives

$$v_{IN} = i_3 \left\{ \left(3 - \frac{1}{\omega^2 C^2 R^2}\right) + \frac{4}{j\omega CR} \right\} \left\{ R + \frac{1}{j\omega C} \right\}$$

$$- i_3 \left\{ 2 + \frac{1}{j\omega CR} \right\} R.$$

From eqn. (4)

$$v_{IN} = v_{OUT} \left\{ \left(3 - \frac{1}{\omega^2 C^2 R^2}\right) + \frac{4}{j\omega CR}\left(1 + \frac{1}{j\omega CR}\right) - \left(2 + \frac{1}{j\omega CR}\right) \right\}$$

$$= v_{OUT} \left\{ \left(1 - \frac{5}{\omega^2 C^2 R^2}\right) + \frac{1}{j\omega CR}\left(6 - \frac{1}{\omega^2 C^2 R^2}\right) \right\}$$

For no " j " term

$$\omega CR = \frac{1}{\sqrt{6}}$$

and

$$f = \frac{1}{2\pi CR\sqrt{6}}.$$

Then

$$v_{IN} = -29\, v_{OUT}.$$

Thus the network gives an output in antiphase to the input but 1/29th of its value, at a frequency given by $f = 1/2\pi CR\sqrt{6}$ Hz. per second.

If the amplifier has a voltage gain of 29 times, the circuit will oscillate at this frequency. In practice the collector load resistance (R_L) should be small and often three stages are required to give sufficient gain. The input impedance of the transistor and biasing network is low enough for appreciable current to be drawn and this must be allowed for in the design of such an oscillator. Typical circuits show that a voltage gain of about 60 times is required between the base and collector of a single-stage oscillator for satisfactory operation.

8.2.2. *Two-stage oscillator with current feedback*

Figure 8.10 shows the circuit of such an oscillator, the dotted lines enclosing the current form of the Wien Network that gives zero phase shift at a particular frequency.

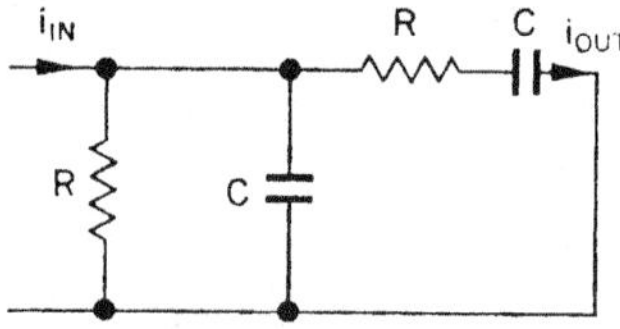

FIG. 8.12. Feedback network for two-stage R.C. oscillator.

The network is reproduced in Fig. 8.12 and

$$\frac{i_{IN}}{i_{OUT}} = \frac{\left(\frac{1}{R}+j\omega C\right)+\frac{1}{R+1/j\omega C}}{\frac{1}{R+1/j\omega C}} = 1+\left(R+\frac{1}{j\omega C}\right)\left(\frac{1}{R}+j\omega C\right)$$

$$= 3+j\omega CR+\frac{1}{j\omega CR}.$$

This shows that at a frequency given by $\omega CR = 1$, i.e. $f = 1/2\pi CR$, the output current is in phase with the input current but one-third of its value.

The complete circuit of the oscillator shows the R.C. network providing the feedback path between the two transistors. If the pair of transistors have a current gain of 3 times, the circuit will oscillate. In practice the current gain is much more than this and must be reduced.

A suitable method is to remove the emitter by-pass capacitor C_E from all or part of the emitter bias resistor R_E. The input resistance is usually not negligible compared with the value of R and for design purposes the series resistance R should be reduced to allow for the input resistance.

8.3. Negative Resistance Oscillators

In this type of oscillator the frequency of oscillation is controlled by a tuned circuit and the losses in the tuned circuit counteracted

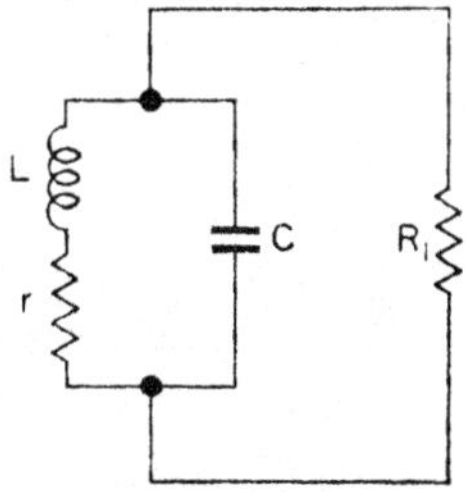

FIG. 8.13. Basic circuit of negative resistance oscillator.

by connecting a negative resistance across it. The circuit shown in Fig. 8.13 will oscillate if R_1 in parallel with R_D is less or equal to zero, where R_D is the dynamic resistance of the tuned circuit,

i.e.
$$\frac{R_1 R_D}{R_1 + R_D} \leqslant 0$$

or
$$\frac{R_D}{1+(R_D/R_1)} \leqslant 0.$$

Since R_D is positive, this condition may only be satisfied if $R_1 \leqslant -R_D$.

Numerically,

if $L = 1$ mH, $C = 100$ pF, $r = 50\ \Omega$, $R_D = L/Cr = 20$ kΩ and if R_1 is -19 kΩ the circuit will oscillate at a frequency

$$f = \frac{1}{2\pi\sqrt{(LC)}} = 159 \text{ kHz} \quad \text{(very nearly).}$$

Negative resistance transistor oscillators are not widely used, but the characteristics of the tunnel diode render it ideally suitable for this purpose.

8.3.1. *Transistor negative resistance oscillator*

Figure 8.14 shows the circuit of a negative resistance transistor oscillator that relies on the phase variation of the current gain with frequency.

As quoted in Table 6.1 the low frequency input impedance of an emitter follower circuit is given by

$$Z_{IN} = r_b + \frac{R_L}{(R_L/r_c)+(1-\alpha_0)}.$$

If R_L is much smaller than r_c,

$$Z_{IN} = r_b + \frac{R_L}{1-\alpha_0} = \frac{R_L}{1-\alpha_0} \quad \text{(very nearly).}$$

At high frequencies the expression for the current gain must be modified to $(\alpha_0/1+jf/f_\alpha)$ and if the emitter load is capacitive, R_L must be replaced by $-jX$,

i.e.

$$Z_{IN} = \frac{-jX}{1-\dfrac{\alpha_0}{1+jf/f_\alpha}}$$

$$= \frac{-jX(1+jf/f_\alpha)}{(1-\alpha_0)+jf/f_\alpha}.$$

If the frequency is such that f/f_α is about 10,

$$Y_{IN} = -\frac{f}{f_\alpha X} + j\frac{(1-\alpha_0)}{X}$$

The input circuit may now be represented by Fig. 8.15, since C_B is only a d.c. blocking capacitor and R_B is very high.

Oscillations will be maintained if

$$-X\frac{f_\alpha}{f} \leqslant -\frac{L}{Cr} \leqslant -\frac{Q}{\omega C} \leqslant -QX,$$

since in this case the input capacity provides the tuning capacitance.

This shows that the circuit will oscillate at a frequency

$$f = \frac{X}{2\pi L(1 - \alpha_0)}$$

if $f \geqslant f_\alpha/Q$.

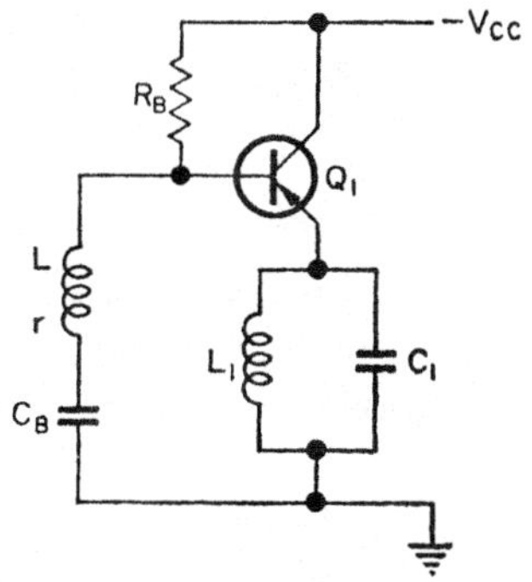

FIG. 8.14. Transistor negative resistance oscillator.

The above analysis is only approximate but indicates the operation of this type of oscillator. In practice a capacity connected in parallel with L would provide a measure of control of the frequency of operation.

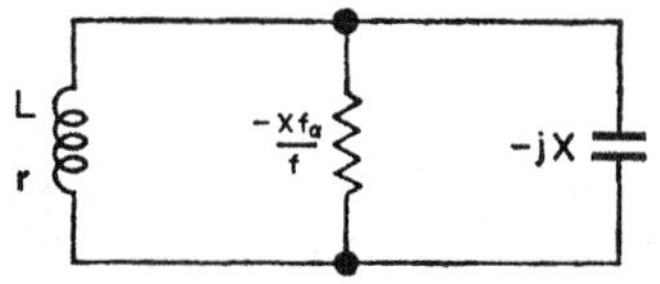

FIG. 8.15. Equivalent circuit of transistor negative resistance oscillator.

The emitter tuned circuit resonates at slightly less than the frequency of operation to provide the capacitive reactance. The use of this tuned circuit enables the d.c. supply current to flow through the coil L_1, and also gives rise to a rapid change of reactance with frequency which ensures better frequency stability.

8.3.2. *Tunnel diode negative resistance oscillator*

This type of negative resistance oscillator uses that part of the characteristic of a tunnel diode where an increase in forward voltage leads to a decrease in current. The characteristic shown and explained in Chapter 2, is reproduced in Fig. 8.16. Over the

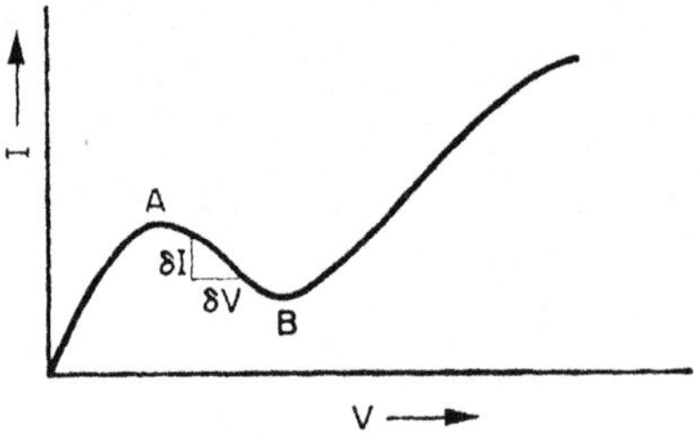

FIG. 8.16. Characteristic of a tunnel diode.

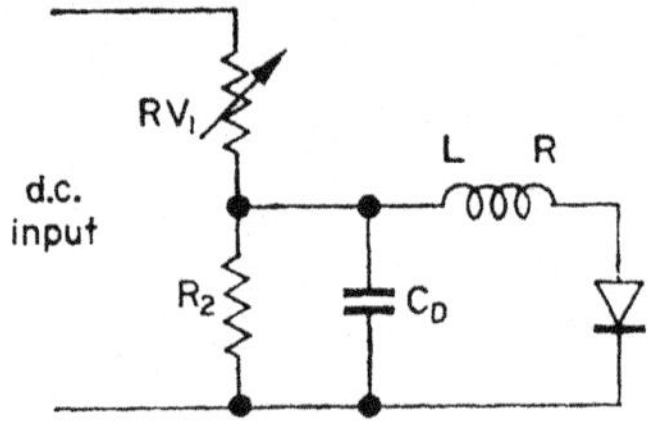

FIG. 8.17. Tunnel diode negative resistance oscillator.

range AB, $\delta V/\delta I$ is negative and if the modulus of this resistance is less than the dynamic impedance of the tuned circuit oscillations will ensue.

The complete oscillator is shown in Fig. 8.17. The variable resistance RV_1 enables the operating voltage of the tunnel diode to be set to a convenient value within the negative resistance range, and the circuit inductance resonates with the junction capacitance of the diode.

The diode may be represented for small signal purposes by the equivalent circuit shown in Fig. 8.18. L_s is the series inductance due to leads and case, R_s is the series resistance due to leads, etc., C is the junction capacitance and $-r$ represents the negative resistance of the diode.

The oscillatory circuit is then as shown in Fig. 8.19 and will oscillate at a frequency

$$f = \frac{1}{2\pi\sqrt{(L_1C)}}, \quad \text{where } L_1 = L+L_s$$

if $r \leqslant -L_1/CR_1$, where $R_1 = R+R_s$.

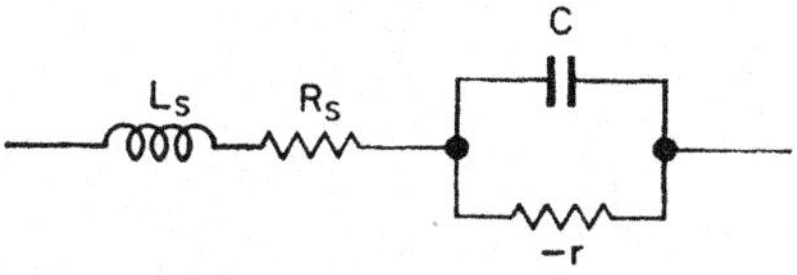

FIG. 8.18. Equivalent circuit of tunnel diode.

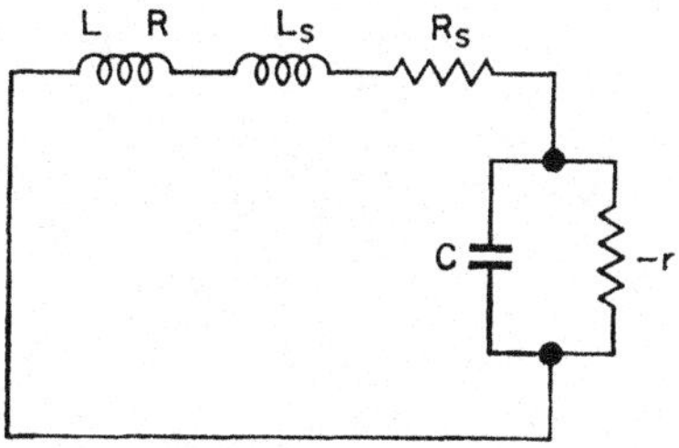

FIG. 8.19. Equivalent circuit of tunnel diode negative resistance oscillator.

Oscillations are limited by the negative resistance range of the diode, but if r is very much less than $-R_D$, the voltage will swing into the positive resistance region and a blocking oscillator will result.

8.4. Crystal Controlled Transistor Oscillators

Circuits using conventional tuned circuits suffer from the variation of inductance and capacitance with temperature.

Changes in ambient temperature also affect the transistor itself. One method of obtaining better frequency stability in a fixed frequency oscillator is to incorporate a crystal in the feedback path or utilize the rapid change of reactance of a crystal with frequency.

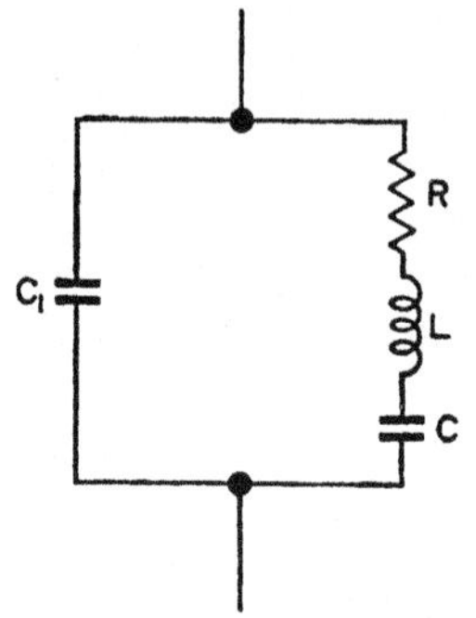

FIG. 8.20. Equivalent circuit of a crystal.

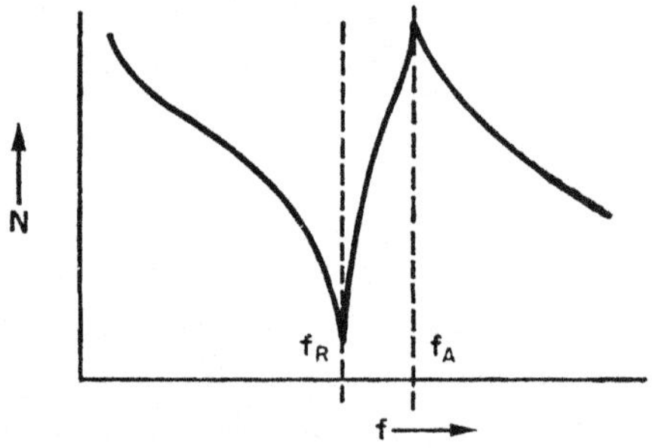

FIG. 8.21. Impedance characteristic of a crystal.

Quartz crystals act as very high Q circuits and rely on mechanical vibrations maintaining electrical oscillations at a fixed frequency. They may be represented by the equivalent circuit shown in Fig. 8.20. This gives rise to an impedance variation as shown in Fig. 8.21 and shows series and parallel resonant frequencies (f_R and f_A) separated by a small frequency range, in which the crystal behaves as an inductance. The smaller the difference between f_R and f_A, the more rapid is the change of inductance with frequency and the higher is the effective Q factor of the crystal.

8.4.1. *Crystal controlled tuned collector oscillator*

The circuit of this fixed frequency oscillator is shown in Fig. 8.22 and is basically the oscillator described in Section 8.1.1. The crystal is in series with the feedback winding and presents a low impedance at the series resonant frequency f_R.

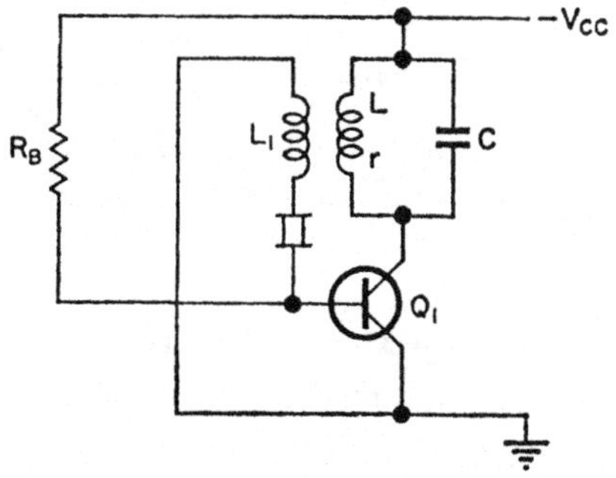

FIG. 8.22. Crystal controlled tuned collector oscillator.

If the collector tuned circuit is resonant at the frequency f_R, the circuit will oscillate and any tendance for the values of circuit components to vary with temperature, supply variation, etc., will be counteracted by the crystal.

8.4.2. *Crystal controlled negative resistance oscillator*

Figure 8.23 shows the circuit of a negative resistance oscillator that uses a crystal operating in the narrow frequency range between the series and parallel resonant frequencies. Over this range the crystal behaves as an inductance and the circuit oscillates in the manner described on Section 8.3.2.

The advantage of this circuit is the rapid change of inductive reactance with frequency as indicated in Fig. 8.21. This will

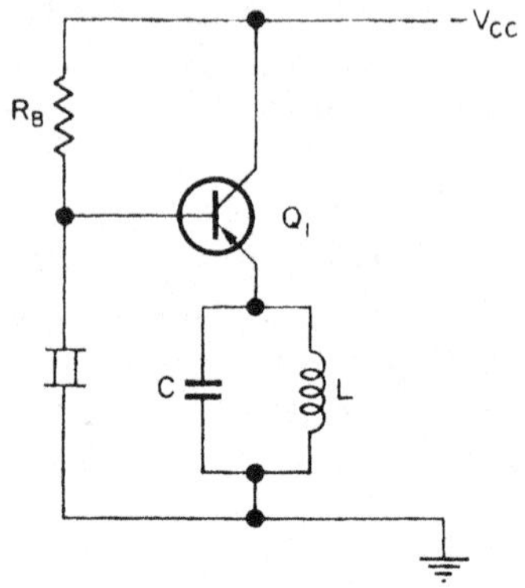

FIG. 8.23. Crystal controlled negative resistance oscillator.

counteract any tendency for the frequency to drift due to change of input capacity caused by supply voltage or temperature variation.

8.5. Inverters and Converters

These applications of transistors are included in this chapter since the oscillator is the most important part of the circuit. Inverters are designed to produce a.c. from a d.c. source, while converters have an additional rectifying stage so that a high or medium direct output voltage is developed. For low power applications, up to 1 kW, transistor inverters and converters compare very favourably with vibrators or rotary converters.

8.5.1. *Ringing choke converter*

Figure 8.24 shows the circuit of a ringing choke converter, which is useful for powers of up to about 5 W. The transistor itself varies between the full-on or bottomed condition and the cut-off state, and a damped oscillation (or ringing) is produced in the secondary winding *ST*. The resonant circuit is composed of the secondary inductance L_s with its associated self and stray capacitances and the diode *D* and smoothing capacitor *C* enable a high direct output voltage to be obtained.

When the transistor is bottomed, virtually all the supply voltage V_{IN} is developed across L_P and the supply or collector current rises approximately linearly since

$$V_{IN} \simeq L_P \frac{dI_c}{dt}.$$

The voltage across the feedback winding EF is approximately constant and the base current is limited by the base resistance R_B. This in turn limits the rising collector current to a value of $\bar{\beta} I_B$, where $\bar{\beta}$ is the large signal current gain. When the supply current

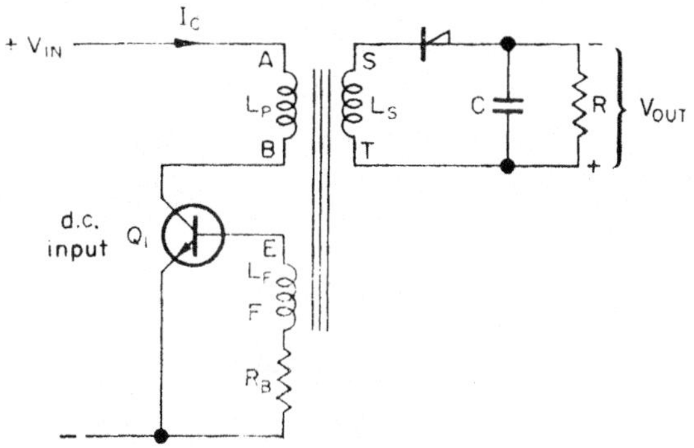

FIG. 8.24. **Ringing choke converter.**

reaches this value, the collector emitter voltage rises, the voltage across L_P falls and the base current falls. The voltages across the windings are reversed and the transistor is cut off, with energy of value $\frac{1}{2}L_P(\bar{\beta} I_B)^2$ joules stored in the magnetic field of the transformer.

This field now collapses, causing ringing in the secondary, and energy is transferred to the output circuit through the diode D. When the current falls to zero, the voltages across the winding reverse, the transistor bottoms and the cycle repeats. Waveforms illustrating the current flow are shown in Fig. 8.25.

The circuit described suffers from bad regulation and the transistor and diode must be able to withstand high inverse voltages. Also a starting bias may have to be applied to initiate oscillations.

8.5.2. *Transformer coupled converters*

This type of converter has a high efficiency (up to 90%) and good regulation, and is suitable for powers within the range 1 W to 1 kW. A typical circuit is shown in Fig. 8.26. Each transistor conducts for half of the time, but the operation does not rely on

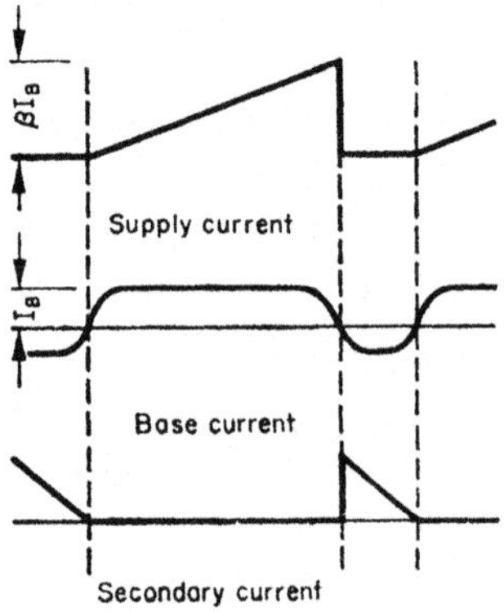

FIG. 8.25. Current waveforms of a ringing choke converter.

the energy stored in the magnetic field as in the previous type of converter. If we consider Q_1 conducting heavily, virtually all the input voltage V_{IN} appears across winding *OB* so that a linearly rising collector current flows, i.e.

$$L_{OB}\frac{dI_{c_1}}{dt} = V_{IN}.$$

This produces an approximately constant base current I_{B_1} due to the feedback winding *GH* and a constant secondary voltage in the winding *ST*. If any secondary current flows, it must be balanced by an increase in primary current I_{c_1}, and under load conditions the total collector current is composed of a constant component and a triangular component. As the collector current increases, the core approaches saturation and the value of the inductance is decreased. This leads to a rapid increase in current until $I_{c_1} = \beta I_{B_1}$ after which I_{c_1} decreases. The fall in I_{c_1} reverses the voltages across the windings and leads to a rapid change over to Q_1 cut-off and Q_2 bottomed. Complete waveforms are shown in Fig. 8.27 and the secondary voltage is approximately a square wave, which may be rectified as explained in

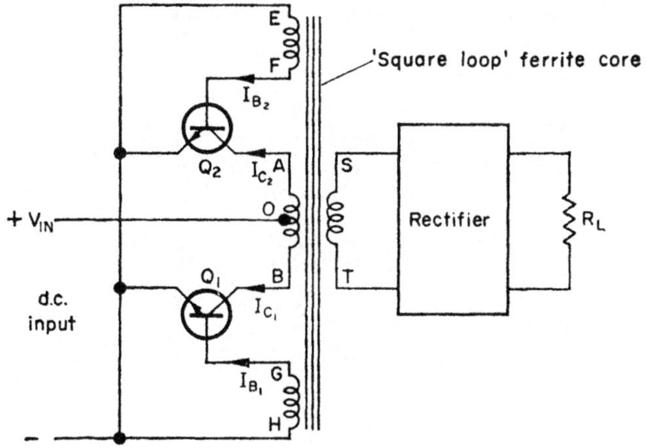

FIG. 8.26. Transformer-coupled converter.

Chapter 5. The frequency of the oscillator of such converters may be derived in terms of the number of turns on half the primary N, the input voltage V_{IN} and the maximum transformer flux Φ_{max},

i.e.
$$V_{IN} = L_{OA}\frac{dI_{c_2}}{dt} = N\frac{d\Phi}{dt}$$

or
$$\frac{V_{IN}}{N}dt = d\Phi.$$

If τ is the time for complete reversal of flux

$$\int_0^{\tau} \left(\frac{V_{IN}}{N}\right) dt = \int_{-\Phi_{max}}^{\Phi_{max}} d\Phi,$$

i.e.

$$\tau = \frac{2N\Phi_{max}}{V_{IN}}.$$

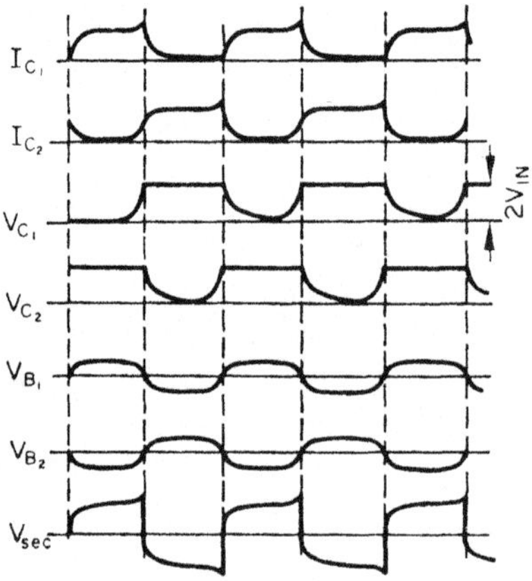

FIG. 8.27. Voltage and current waveforms of a transformer-coupled converter.

The operating frequency is then

$$f = \frac{1}{2\tau} = \frac{V_{IN}}{4N\Phi_{max}}.$$

In practice this frequency usually lies within the range 100–1500 Hz with 400 Hz being a common design value.

Starting circuits are often required and probably the simplest is shown in Fig. 8.28. Initially, a small current will flow through R into the base connections of both transistors and the one with the higher value of current gain will conduct first. As soon as the converter starts operating, however, the base current flows through the diode and the power loss in the resistor is negligible.

Germanium transistors are usually used for such circuits due to their low collector saturation resistance, 0·05 Ω compared with 5 Ω for silicon, but they are limited by peak collector currents and voltages.

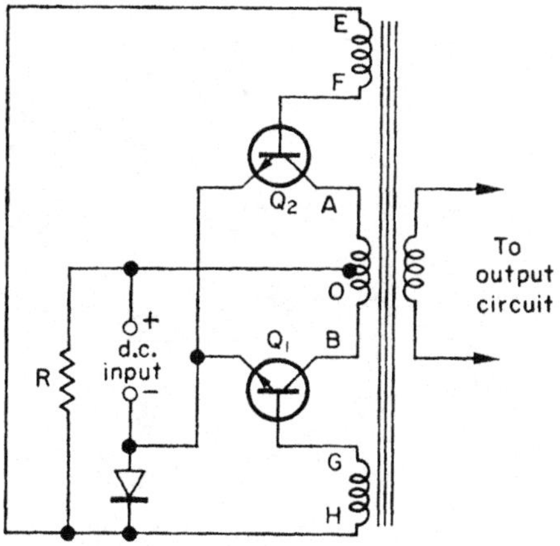

FIG. 8.28. Transformer-coupled converter showing starting circuit.

Questions for Chapter 8

Q.8.1. Show that, if sinusoidal operation is assumed, the Colpitts type oscillator (Fig. 8.3) oscillates at a frequency given by

$$\omega^2 = \frac{1}{LC}\left[1 + \frac{r(1-\alpha_0)}{R}\cdot\frac{C}{C_1}\right],$$

where

$$\frac{1}{C} = \frac{1}{C_1} + \frac{1}{C_2}, \quad R = r_e + r_b(1-\alpha_0).$$

Determine also the condition for oscillations to be initiated. The effect of the bias circuit and r.f. choke may be neglected.

Q.8.2. Show that the circuit given in Fig. 8.29 will oscillate if

$$M = \frac{RrC}{\alpha_0} + \frac{L_1}{\alpha_0},$$

where
$$R = r_e + r_b(1 - \alpha_0).$$

Determine the frequency of oscillation of the circuit. Assume sinusoidal operation and ignore the effects of the bias circuit.

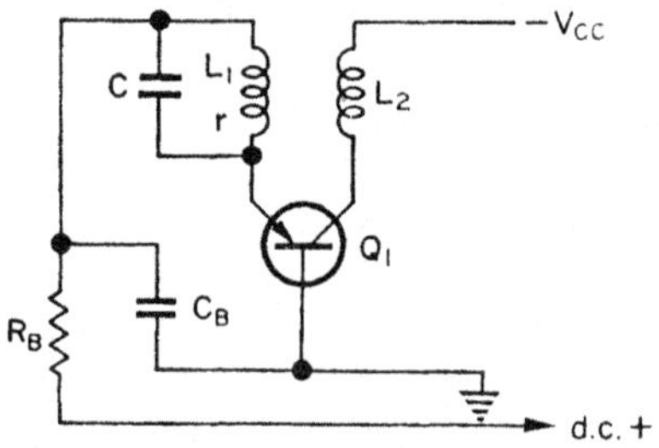

FIG. 8.29. Circuit for Q.8.2.

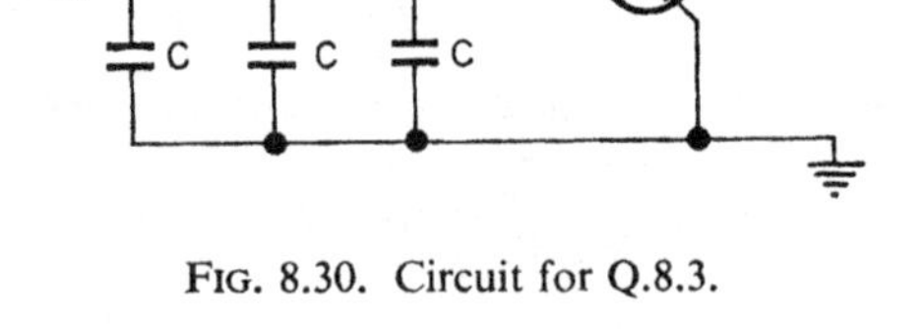

FIG. 8.30. Circuit for Q.8.3.

Q.8.3. Derive an expression for the approximate frequency of oscillation for the R.C. oscillator shown in Fig. 8.30. Assume the load and bias resistors (R_L and R_B) are very high, the input resistance is low and the reactance of the coupling capacity C_c may be ignored.

Evaluate when $R = 2{\cdot}2\ \text{k}\Omega$, $C = 0{\cdot}01\ \mu\text{F}$.

Q.8.4. Show that the current gain of a grounded base transistor is approximately equal to the current amplification factor α if the impedance connected between the collector and the base is much less than the collector resistance r_c. In the circuit of Fig. 8.31 the reactance of the capacitor $-X_c$ is much less than r_c at the frequencies of interest.

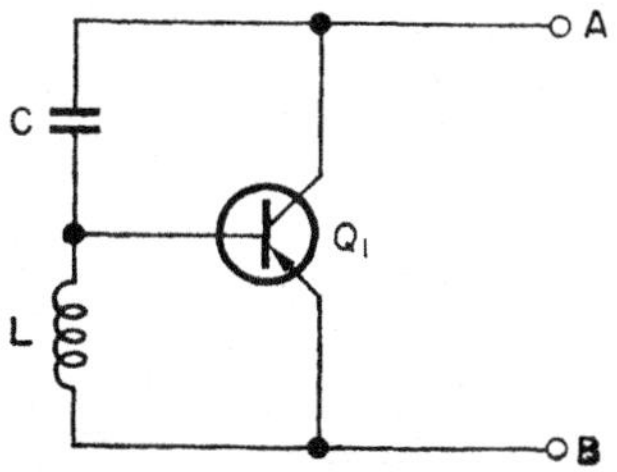

FIG. 8.31. Circuit for Q.8.4.

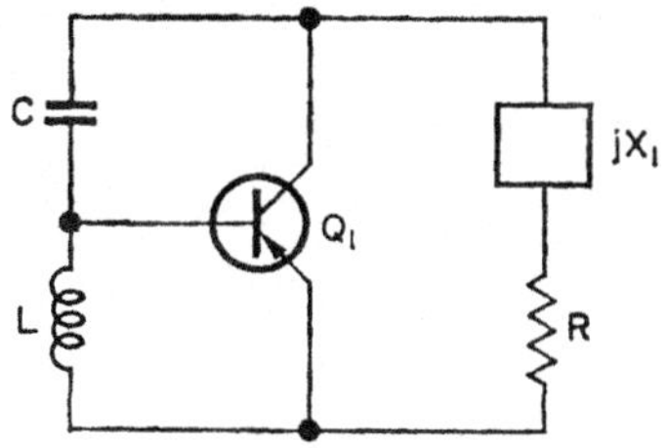

FIG. 8.32. Circuit for Q.8.4.

If X_L is the reactance of the inductor show that the condition X_L/X_c is very much less than α must be approximately satisfied if the input impedance at the terminals AB has a negative resistance part. Hence determine the nature and magnitude of the reactance X_1 in Fig. 8.32 to cause oscillation at a frequency for which the reactances of the inductor and the capacitor are 500 Ω and −625 Ω respectively. What is the maximum value of R for which the circuit will oscillate? Any reasonable approximations may be used.

The parameters of the transistor in Fig. 8.32 are:

$$r_e = 50\ \Omega,\ r_b = 500\ \Omega,\ r_c = 500\ \text{k}\Omega,\ \alpha = 0{\cdot}98. \qquad \text{(B.Sc., 1961)}$$

Q.8.5. Sketch the characteristics of an Esaki or tunnel diode indicating approximate scales of voltage and current. Explain briefly its importance as a circuit element.

Figure 8.33 indicates a possible equivalent circuit for the tunnel diode. Derive an expression for the input impedance and determine the frequencies for which the real and quadrature components become zero if $R = 1{\cdot}5\ \Omega$, $L = 0{\cdot}01\ \mu\text{H}$, $C = 50\ \text{pF}$, $r = -25\ \Omega$. What is the significance of the results? (I.E.E., June 1963)

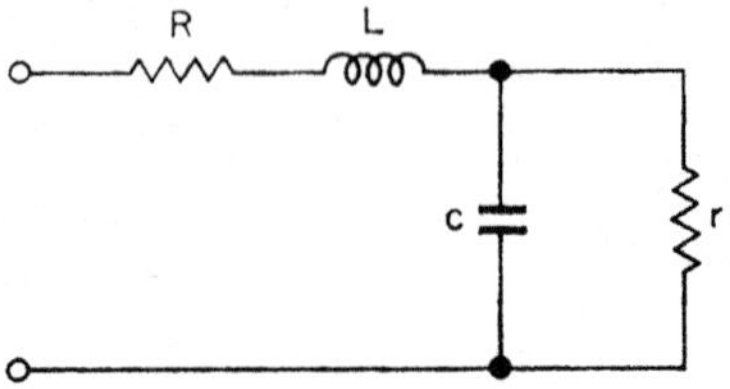

FIG. 8.33. Circuit for Q.8.5.

CHAPTER 9

Pulse and Computing Circuits

9.1. Introduction

In this chapter we are concerned with non-sinusoidal operation of semiconductor devices. The transistor is often used in pulse or switching circuits in such a manner that it is driven from the cut-off region to the bottomed region (or vice-versa) without remaining for much time in the normal, active, region of operation. In these circuits the silicon transistor has a particular advantage of low cut-off current and the germanium transistor has low bottoming (or saturation voltage). These give rise to high efficiency and well-defined ON and OFF states.

An important point of circuit design is the speed with which the transistor can be switched through the normal region of the characteristics. During the switching process a phenomenon known as " charge carrier storage ", which is hole storage in the case of *pnp* transistors, occurs. In order to obtain transistor action it is necessary to establish and maintain a certain charge of carriers in the base region. This has given rise to the " charge-controlled " approach to switching circuit design, which was first suggested by Beaufoy and Sparkes. By considering current to be the result of the flow of charge into and out of a base store, a possible approach to small and large signal calculation is obtained, which may be helpful in the design of some of the following circuits. However, we shall not study this design procedure in detail.

9.2. Astable Multivibrator

The free-running multivibrator circuit is essentially a resistance–capacitance coupled two-stage amplifier, with a large amount of positive feedback, such that it oscillates to give an output waveform which is rich in harmonics of the fundamental frequency. The basic circuit is shown in Fig. 9.1, and is a valuable source of square or rectangular waves, with simple control of both the frequency and the mark : space ratio.

When the d.c. supply is switched on the transistors will go into one of two possible unstable states, due to slight unbalance in the components, or random variations, such as electrical noise, in the

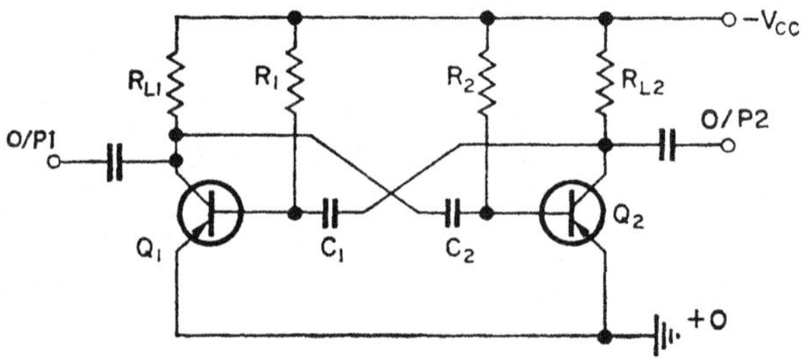

FIG. 9.1. Basic astable multivibrator circuit.

voltages. The circuit is self-starting with one transistor going ON, into conduction, and the other transistor going OFF, with only leakage current flowing.

If we can assume an initial condition where Q_1 is ON and Q_2 is OFF, then as C_2 charges through R_2, so the base voltage of Q_2 goes negative and ultimately transistor Q_2 comes into conduction. At this time regenerative switching occurs, with the collector voltage of Q_2 going quickly positive, due to current flow through the load resistor R_{L2}. This gives rise to a positive pulse, through capacitor C_1 to the base of Q_1 and thus turns that transistor OFF. Capacitor C_1 then charges with an approximate time constant of C_1R_1, until the base of Q_1 is sufficiently negative to the emitter for Q_1 to come ON again, and for the regenerative cycle to repeat.

The load resistance is chosen to limit the collector current to a suitable value for the transistor in the circuit. Then, to ensure bottoming, the base resistor must not be too large to limit the base current. In other words

$$R_1 < \beta_1 R_{L1} \quad \text{and} \quad R_2 < \beta_2 R_{L2},$$

where the current gain values will be the lowest expected for the type of transistor used. Unequal ON and OFF times can be achieved by making the time constants C_1R_1 and C_2R_2 unequal, but this simple circuit is not suitable for mark : space ratios much greater than 10 : 1.

The width of the output pulse corresponds to the time that either Q_1 or Q_2 remains OFF. This time may be estimated by ignoring the leakage current. When transistor Q_1 is ON, the base to earth voltage, corresponding to saturation, is V_{BS} and at the same time Q_2 is OFF, with a collector to earth voltage of V_{CC}. After the transition the collector voltage of Q_2 goes to the bottoming voltage of V_{CS} and thus a voltage pulse of $V_P = V_{CC} - V_{CS}$ is transmitted through C_1 to the base of Q_1, turning that transistor OFF. Transistor Q_1 will be turned ON again when the base voltage has recovered exponentially to a value of V_{BO}. Hence the time that Q_1 is OFF is given by

$$t_1 = C_1R_1 \log_e\left[1 + \frac{V_{CC} - V_{CS} - (V_{BS} - V_{BO})}{V_{CC}}\right] \text{ seconds.}$$

The value for t_2 may be calculated in the same way, so that the periodic time for a complete cycle of multivibrator operation is given by

$$T = t_1 + t_2.$$

The approximate base and collector voltage waveforms are shown in Fig. 9.2. The collector waveform may be improved by adding a clipping diode (joined to a voltage less negative than $-V_{CC}$) at each output point. In fact for a small germanium transistor the difference between V_{BS} and V_{BO} is very small, and

V_{CS} is much less than V_{CC}. Ignoring these small values, we have that:

$$t_1 = 0{\cdot}69C_1R_1 \quad \text{and} \quad t_2 = 0{\cdot}69C_2R_2.$$

While a transistor is OFF, and at any given temperature, the leakage current is constant, and provides a second discharge path for the capacitors C_1 and C_2. The effect of leakage is thus to reduce the periodic time, and this effect becomes more marked with higher temperatures.

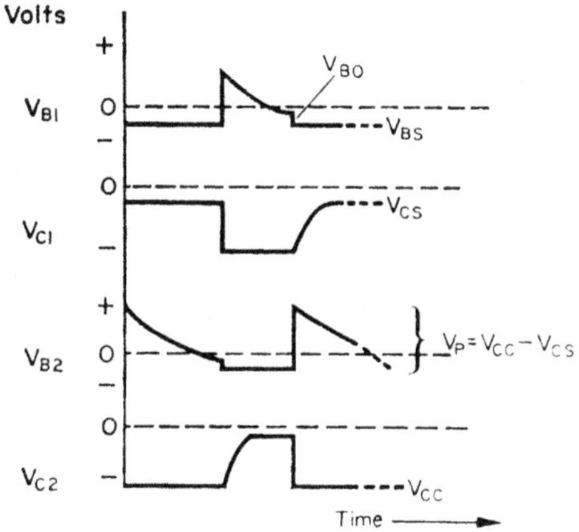

FIG. 9.2. Voltage waveforms in the astable multivibrator.

The limit to the pulse repetition frequency of any multivibrator circuit is set by switching times of the transistor. The circuits given in this chapter operate satisfactorily to a frequency about 20% of the current gain cut-off frequency in common base (f_α).

9.3. Monostable Multivibrator

This form of multivibrator, sometimes known as the flip-flop, has one stable and the other unstable, or quasi-stable, state. The circuit, with *pnp* transistors is given in Fig. 9.3, where the positive bias voltage V_B is 1 or 2 V. A trigger pulse switches transistor Q_1 OFF, which " flips " the circuit into the unstable state. Due to the capacitive cross-coupling the circuit subsequently " flops " back into the stable state.

This monostable circuit differs from the astable multivibrator in that one capacitive a.c. cross-coupling is replaced by a resistive d.c. coupling, and the base of one transistor is taken to a bias voltage, to hold that transistor OFF in the normal condition. The

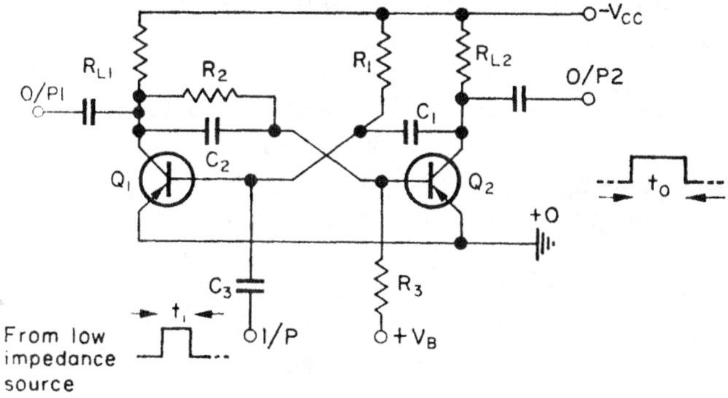

FIG. 9.3. Monostable multivibrator circuit.

small capacitor C_2 known as the " speed-up " capacitor, is included to give a rapid transfer of charge into the base of Q_0 and thus reduce the switching time.

The time length t_i of the input pulse need only exceed the required voltage for the time necessary for the transistors to switch over, while the duration of the output pulse depends mainly on the time constant C_1R_1 and so has a time length of

$$t_0 \simeq 0{\cdot}7C_1R_1.$$

The minimum input level to ensure switch-over is the sum of the bias voltage V_B (for which 1·5 V is often convenient) and the largest value of base to earth voltage for a transistor to conduct, which is usually less than 0·5 V. In the stable state Q_1 is bottomed and Q_2 is OFF. A positive trigger-pulse to the base of Q_1 turns that transistor OFF and thus the collector voltage of Q_1 goes negative, to overcome the bias voltage V_B and turn Q_2 ON. Subsequently, capacitor C_1 discharges through R_1, and also by leakage current through the transistor, and transistor Q_1 comes ON again when the base to earth voltage is slightly negative. Since the values of the resistors are more or less fixed by d.c. conditions, any wide variation of the output pulse width t_0 must be controlled by the capacitor C_1.

Some applications of the monostable multivibrator circuit are as follows:

(a) Pulse Shaping—a short non-rectangular pulse may be shaped into a rectangle of controllable time length.
(b) Pulse Amplification—an input pulse of 1 or 2 V is amplified to almost the full supply voltage V_{CC}, with the possibility of either negative or positive going pulses from the collectors of Q_1 or Q_2.
(c) Pulse Delay—positive or negative going voltages are available after the variable time period t_0.

9.4. Bistable Multivibrator

The circuit shown in Fig. 9.4 has two stable states, with resistive cross-couplings between both transistors. If one transistor is ON, a positive pulse to that base will turn the transistor OFF, and so bring the other transistor ON. Providing that the gain through the circuit, from base to base, is greater than one, this switch-over is almost immediate.

The " steering diodes " (D_1 and D_2) ensure that the input pulse is directed to the base of the transistor which is ON. For example, when transistor Q_1 is ON, diode D_1 has practically zero bias voltage and is on the point of conducting. At the same time diode D_2 will have a negative bias almost equal to the supply voltage V_{CC}. Under this condition the next positive input pulse goes to the base of Q_1 and switches that transistor OFF. For every two input pulses there will be one output pulse from either collector. This circuit is therefore valuable as a " divide-by-two ", this being the basis of binary counting, and also one form of electronic " memory ".

The time constant of the input circuit (C_3R_5 or C_4R_6) sets the limit for the time width of the input pulse. Between input pulses there must be time for the input capacitor to return to nearly the initial voltage condition. For effective triggering this time constant

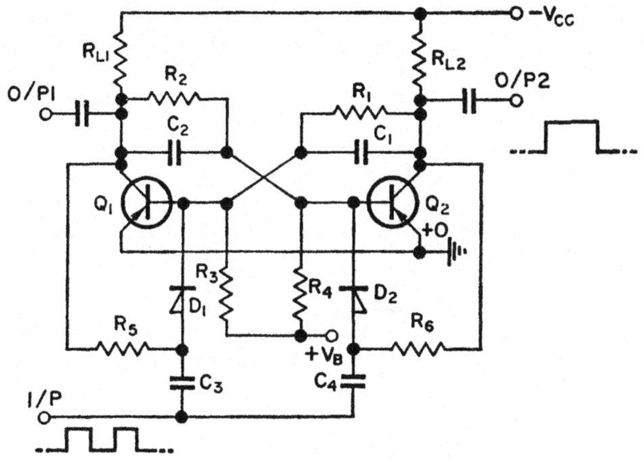

FIG. 9.4. Symmetrical bistable multivibrator circuit.

needs to be about five times the width of the trigger pulse. The maximum pulse counting rate is achieved with careful design and selection of high frequency transistors, as with the other multivibrator circuits. The highest speeds of operation are obtained by coupling with a transformer, rather than by a capacitor, from the output collector.

By using *n* binary stages in series, with resistance–capacitance coupling, the original pulse train may be divided by 2^n times and the state (ON or OFF) of the various transistors at the end of a count can indicate the number of pulses received.

Another interesting bistable multivibrator circuit uses a complementary pair of transistors, one *pnp* and one *npn*. This neat arrangement is illustrated in Fig. 9.5. The output from the collector of a *pnp* transistor is of correct polarity for direct coupling to the base of an *npn* transistor. The current gain of such a combination will be greater than one to a high frequency.

In this circuit both transistors will be either OFF or ON simultaneously, and with the proper choice of load resistors two stable states are possible. The circuit is sensitive to either negative or positive going input pulses. By joining the base of the *npn* transistor Q_2 to a voltage more negative than its collector supply it is possible for the base current to reverse, thus improving the switch-off time.

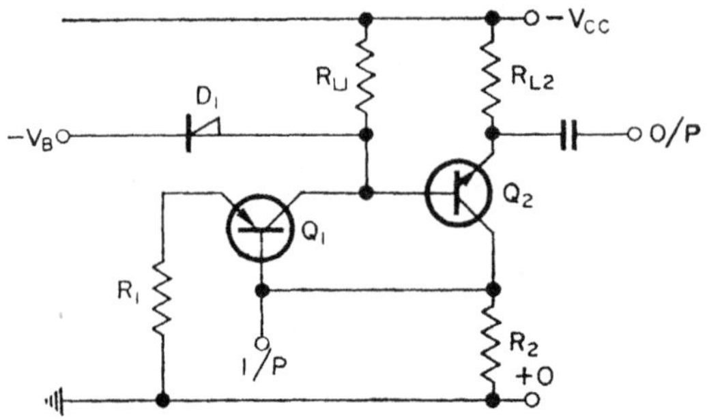

FIG. 9.5. Complementary bistable circuit.

The transistor pair could be replaced by one *pnpn* transistor, which is effectively a low-power controlled rectifier. This device is discussed more fully in Section 9.6.

9.5. Binary and Decade Counters

There are several ways in which individual bistable stages may be interconnected to operate as pulse counting circuits, either to the binary or decimal scale. A binary counting arrangement is given in the block schematic of Fig. 9.6. Each of the four blocks represents a complete divide-by-two circuit, such as that of Fig. 9.4. The 0 or 1 represents the normal OFF or ON condition of the transistors, after reset has been applied, but before any input pulse has been received. An indicator, which is shown as a lamp in this example, may be joined to the normally OFF side of each binary stage, and the state of the indicators at any time shows the number of pulses received, up to the maximum number possible for the circuit (15 in this case). Capacitor interconnection, with a pair of steering diodes, is required between each stage. The collector voltage swing from the first stage is differentiated and supplies the pulse needed to drive the next stage.

A scale of ten is frequently required for decimal counting and may be achieved by suitable pulse feedback arrangements between the binary stages. One possibility is to have feedback from the

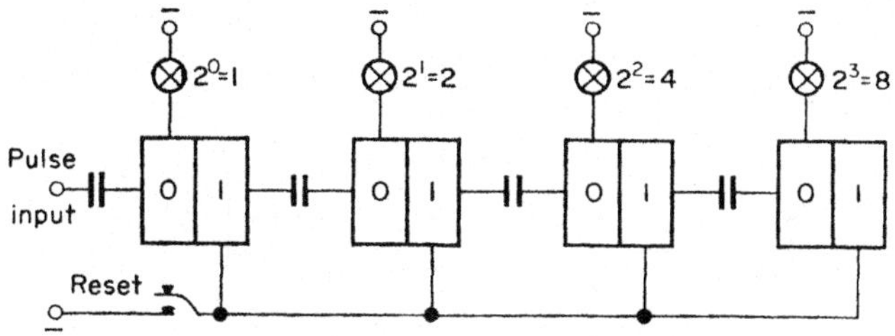

FIG. 9.6. Four-stage binary counter.

fourth to the third and from the third to the second stages. Four binary stages would normally divide by 2^4, that is, count 16 pulses and give one output pulse on the sixteenth input pulse. However, the setting of the third stage may be used to feed an extra pulse into the second stage. This adds two to the count. Then the setting of the fourth stage feeds an extra pulse into the third stage. This adds four. So the feedback loops add six each cycle and an output pulse will be obtained after ten input pulses. A slightly different arrangement is to have simultaneous feedback from the fourth to both third and second stages. This is illustrated in outline in Fig. 9.7, and may be understood from the table of Fig. 9.8. The feedback arrangement ensures that when the fourth binary stage changes to the ON state this makes the third and second stages follow. As described previously each stage is reset such that the output transistor is ON in order that the counting starts correctly at 0.

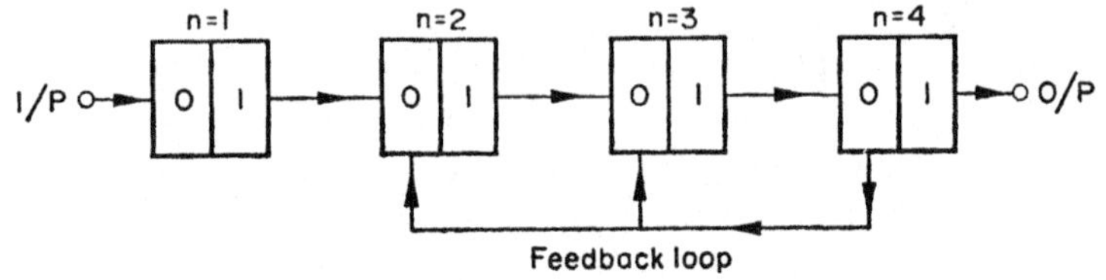

FIG. 9.7. Decade counter using feedback.

Input pulse	Binary stage				Output pulse
	$n=1$	$n=2$	$n=3$	$n=4$	
0	0	0	0	0	
1	1	0	0	0	
2	0	1	0	0	
3	1	1	0	0	
4	0	0	1	0	
5	1	0	1	0	
6	0	1	1	0	
7	1	1	1	0	
8	0	1	Feedback 1	1	
9	1	1	1	1	
10	0	0	0	0	1

FIG. 9.8. Pulse-counting sequence to add 6 counts.

Theoretically, binary stages may be used with the appropriate feedback arrangements to give any desired scale (such as 12 or 20). However, in practice these pulse feedback circuits present difficult design problems. One difficulty is that the feedback pulse must be interleaved between two incoming pulses, since any overlap of the leading or trailing edges would mean the loss of a pulse.

Most decade counting circuits employing transistors and feedback between binary stages do not operate at frequencies much greater than 1 MHz.

9.6. Ring Counter with *pnpn* Transistors

The ring counting configuration needs one bistable stage per pulse count, and therefore appears to be wasteful of components.

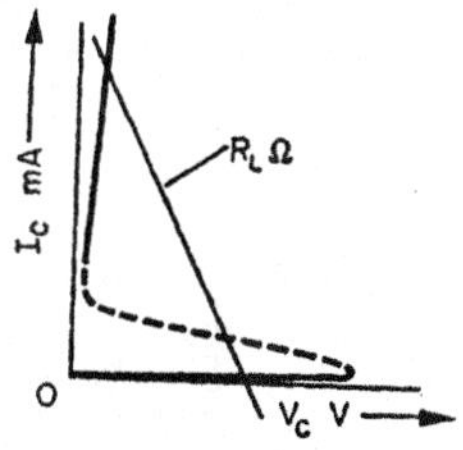

FIG. 9.9. Collector characteristic of *pnpn* transistor.

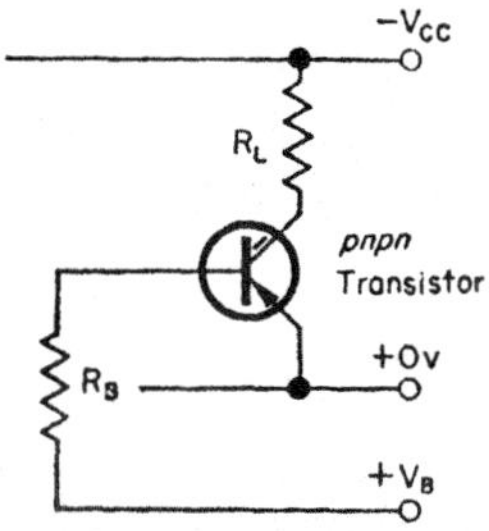

FIG. 9.10. Test circuit for *pnpn* transistor.

However, the arrangement does have some definite advantages. The ring may be made to any required scale size (such as 7, 10 or 12) without the use of feedback, which is often difficult to make reliable. All the stages are identical, and direct indication of the state of the count is possible without auxiliary binary to decimal conversion, which is required with the decade circuits of Section 9.5.

The four-layer transistor, with three electrodes, is very suitable for use in a ring configuration circuit since it is inherently a bistable device with a thyratron-like characteristic. The germanium *pnpn* transistor is best described by the collector current/voltage characteristic of Fig. 9.9. The transistor is normally OFF, over a wide range of collector voltage. If the transistor is tested in the simple circuit of Fig. 9.10 the leakage current flows in the base resistor R_B.

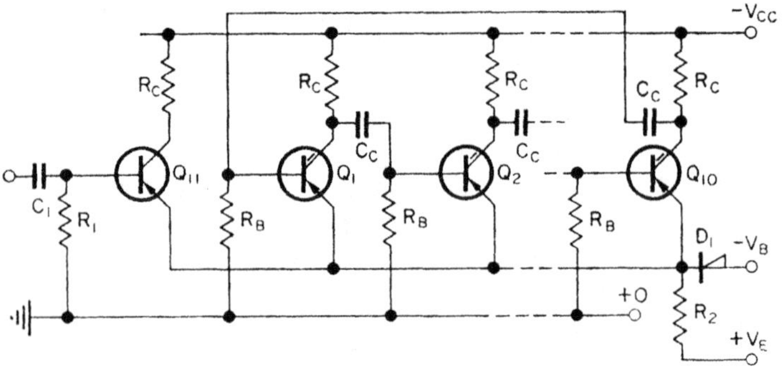

FIG. 9.11. Decade counting ring circuit with *pnpn* transistors.

As the collector voltage is increased the leakage current increases to the point at which the base to emitter voltage passes through zero, and the transistor then conducts. As the current gain between emitter and collector is greater than one (as explained in Chapter 5 for the controlled rectifier) the transistor is in an unstable condition and promptly switches to the bottoming point, determined by the collector load. The *pnpn* transistor is thus a device which is either OFF, with very little current flowing, or bottomed, with an internal volt drop of less than 1 V. The transistor is turned OFF again by reducing the collector or emitter current to zero.

Ten *pnpn* transistors may be joined in series, as shown in Fig. 9.11, to make a convenient decade ring counter. Using available germanium *pnpn* transistors this circuit will count pulses reliably to a frequency of 100 kHz.

With the emitters of all the transistors of Fig. 9.11 in common, the emitter current is defined by the resistor R_2 to the $+V_E$ volts line and the catching diode to $-V_B$ volts. This emitter current could flow in any transistor, but the resistors and supply voltages are chosen so that one transistor in the bottomed condition takes all the available current. Thus only one transistor of the ring can be ON at any one time, and the other transistors will then be cut off.

When the *pnp* driving transistor Q_{11} is turned ON by a negative going input pulse, of about -5 V, this diverts all emitter current from the ring. The transistors Q_1 through to Q_{10} will be OFF. The *pnpn* transistor that was previously ON passes a pulse, via the coupling capacitor C_C to the base of the next transistor. This primes that transistor so that, when the input pulse disappears, and the parallel driving transistor Q_{11} goes OFF, then the primed transistor comes ON. Clearly, there is a low frequency limit to the operation, depending on the time constant associated with the coupling capacitor. This time constant, in turn, depends on the product $C_C(R_C+R_B)$ and the leakage current through the *pnpn* transistor. For each successive input pulse the ON state progresses round the transistor ring, to give one output pulse for every ten negative pulses applied to the input.

9.7. Digital Computing Operations

In an electronic computer numbers and other items of information are represented as the values of sets of electrical quantities. The circuit then operates on these quantities, according to mathematical rules, to give a set of values which corresponds to the resultant of the computation.

An electrical quantity can represent a numerical digit by taking a value which is either within a continuum of values, or a value which is one of a discrete set of values. The first method is that of analogue computation, which is not considered in this chapter.

We are here concerned only with the circuits associated with the second, or digital, method of representation. Most digital computers code all the information in the binary, or scale-of-two form, as there are then only two possible digits, 0 and 1, and the mathematical rules are particularly simple. The use of both digital representation and the binary scale make the translation of mathematical operations into circuitry a fairly straightforward procedure.

The mathematical rules are, in general, a series of logical operations or decisions. It is found that all logical operations can be considered as multiple applications of the elementary operations of either AND or OR and INVERSION.

Firstly, a digit x is the result of the AND operation on digits a, b, c, ..., if

$$x = 0 \text{ when any of } a, b, c, \ldots = 0$$

and $$x = 1 \text{ only when } \textit{all} \text{ of } a, b, c, \ldots = 1.$$

Secondly, a digit x is the result of the OR operation on the digits a, b, c, ..., if

$$x = 0 \text{ when all of } a, b, c, \ldots = 0$$

and $$x = 1 \text{ when one or more of } a, b, c, \ldots = 1.$$

Thirdly, a digit x is the result of the INVERSION operation on digit a, if $$x = 0 \text{ when } a = 1$$

and $$x = 1 \text{ when } a = 0.$$

To make an electronic digital computer, it is necessary to represent numbers and other information as electrical signals, to operate on these signals with circuits to give the AND, OR and INVERSION functions, to store the digits for short or long periods and to convert the resultant waveform into a meaningful output.

In the digital computer the digits are normally represented by pulses, the time length of the pulse determining the maximum speed of computation. One possible arrangement employs a relatively positive voltage to represent 1 and a relatively negative voltage to represent 0. In this system the logical information is contained in the steady-state voltage or current in the circuit. Another possibility is for the information to be contained in the presence (1) or absence (0) of a pulse at a particular time, according to a central clock.

In the remainder of this chapter we are concerned with a few systems of semiconductor circuit logic which are widely used. Semiconductor diodes and transistors have a number of characteristics which are desirable in a computer. Devices which are small, light and low powered are essential, when even a small computer may contain several thousand AND, OR and INVERSION circuits. The low bottoming voltage means that a germanium transistor approaches an ideal switch, whilst the availability of both *pnp* and *npn* transistors allows elegant circuits to be devised. However, the circuits must be designed to allow for initial manufacturing tolerances and any subsequent drift during life. For example, the common emitter current gain β of a transistor has a large initial tolerance and also changes with both temperature and time. For these reasons some method other than base current should be used to define the collector current in these switching circuits.

9.8. Diode Gate Circuits

The performance of AND and OR functions with semiconductor diodes is quite fast and economical, and is therefore popular. Figures 9.12 and 9.13 show AND and OR gates respectively, where the convention being used is that the relatively negative input voltage V_0 represents 0 and the relatively positive voltage V_1 represents 1.

For the AND circuit the output signal is 1 when input signals of 1 are applied to input *A* AND *B* AND *C*. In the case of the OR circuit the output signal is 1 if an input of 1 is applied to *A* OR *B* OR *C*. Successful operation requires that the input is fed from a low resistance source to complete a d.c. loop and the output is put into a high impedance circuit which does not take any significant current from the supply.

The operation of these gates is most easily understood if the diodes are considered to be ideal, with no forward volt drop and no reverse current. Then the AND gate always gives an output equal to the least, or most negative, input voltage and the OR gate output corresponds to the highest, or most positive, input voltage.

Note that if the convention about 1 and 0 is reversed, then the roles of the AND and OR gates are interchanged. So Fig. 9.12 may

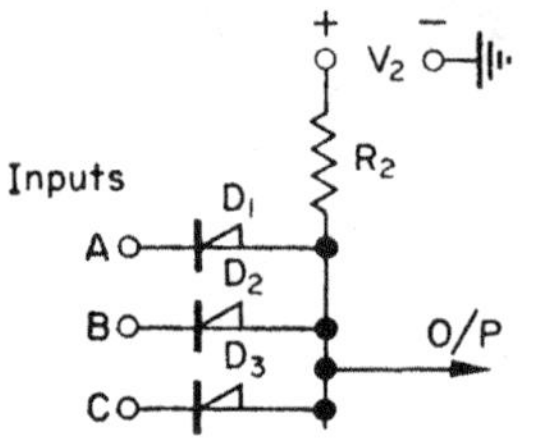

FIG. 9.12. Diode AND gate with three inputs.

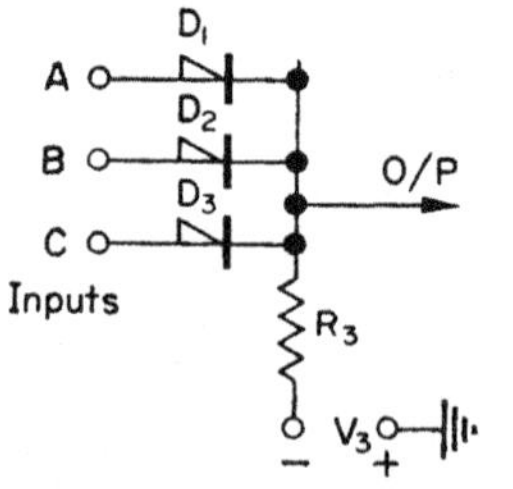

FIG. 9.13. Diode OR gate with three inputs.

be labelled as a positive AND or a negative OR gate, thus allowing for 1 to be either a positive or negative voltage.

The computer logic often requires a combination of AND and OR gates, with the output of one circuit acting as the input to the next. The circuit is then said to have two switching levels and this introduces certain problems additional to those of single-level circuits. A simple example is given in Fig. 9.14 when two AND gates, each with two inputs, feed one OR gate. The output will be V_1 if A AND B OR C AND D are V_1, where

$$+V_2 > V_1 > V_0 > -V_3.$$

The relationship between the supply voltages and resistance values (R_2 and R_3) is critical and the resistors are usually calculated by working back from the output, assuming that the load resistance and the diode characteristics are known. A two-level

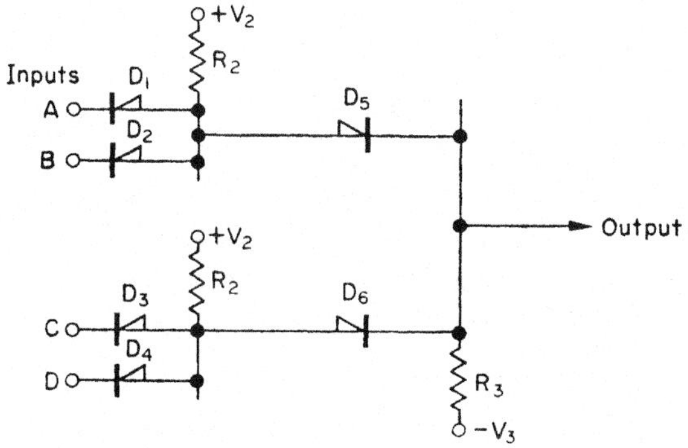

FIG. 9.14. Two-level AND to OR gate circuit.

AND to AND gate circuit is changed by a trivial rearrangement to a single level AND gate, with fewer components, and need not be considered.

Three- and four-level diode gates are theoretically possible, but it is not usual to have more than two levels in tandem without an intermediate amplifier. One difficulty is that, when proceeding from output to input in the design each resistor must draw current which overcomes the effect of the previous resistor, and this results in excessive current drain and possible overload of the diodes.

The small point-contact diodes which are usually employed in gate circuits have several non-ideal characteristics which need to be taken into account. The diodes have finite reverse resistance which means that some current is always being drawn and imposing a load on the input circuit. This effect would be pronounced with a two-level gate having many inputs.

Non-zero forward resistance and forward starting voltage give a volt drop across the diodes. So each stage of diode gating causes a power loss and signal deterioration. An emitter follower amplifier is useful between successive diode stages since it provides power gain and the desirable conditions of high input and low output impedance.

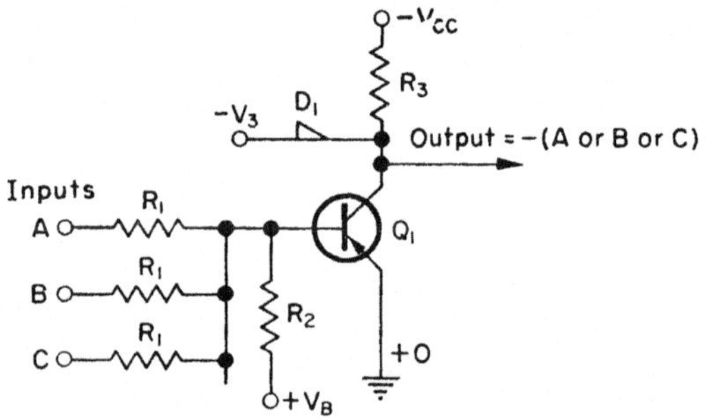

FIG. 9.15. Transistor–resistor NOR circuit.

Diode junction capacitance, which is usually less with point contact than with *pn* junction diodes, limits the rise and fall times of the voltage pulses. This capacitance varies with change of d.c. conditions. Charge carrier storage, sometimes known as " Hole Storage ", means that the diodes take a finite time to turn ON and OFF. This leads to positive and negative going " spikes " on the output waveform.

9.9. Transistor–Resistor Logic

An OR circuit which gives an inverted output, and is therefore sometimes known as a NOR unit, is given in Fig. 9.15. Inputs A, B and C are voltage sources which may take one of the two values V_0, which is approximately zero, and $-V_1$. The resistance values are chosen such that the transistor conducts and bottoms when one or more of the input voltages is at $-V_1$, and then the output voltage changes from $-V_3$ to near zero. The output of one stage may be joined directly as the input of another transistor–resistor logic circuit. However, this type of circuit is slow in operation. In one extreme the transistor may be turned ON with a small base current of

$$I_{\text{MIN}} = \frac{V_1}{R_1} - \frac{V_B}{R_2}.$$

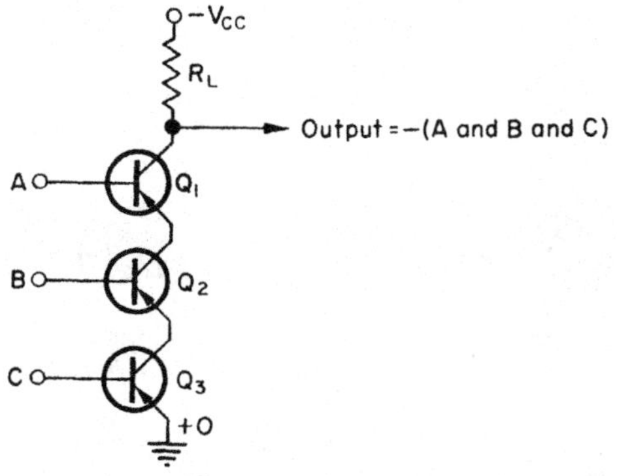

FIG. 9.16. Three-transistor NAND circuit.

This results in a slow rise time. In the other extreme all the inputs, possibly up to ten in number, may be at $-V_1$. This will cause a large base current to flow, and there will be a long turn-off delay when all the inputs change to zero potential.

Desirable qualities for the transistors to be used in these logic circuits are low bottoming voltage, narrow spread in current gain and short charge carrier storage time.

9.10. All-transistor Switching

This form of logic circuit uses one transistor per input. Figure 9.16 shows a three-input AND gate, which also provides an inverted output voltage.

The inputs A, B and C are voltage sources, where a relatively positive voltage V_0 represents 0 and $-V_1$ is equivalent to 1. When any one of the transistor bases is at $+V_0$ then only leakage current flows through the resistor R_L and the output voltage is at almost $-V_2$. Only when all the bases are at $-V_1$ do the transistors conduct and saturate to give a near-zero output voltage.

This type of circuit gives a voltage output which is unsymmetrical about earth and is therefore most unsuitable for driving directly into the input of a similar circuit. Again, although this

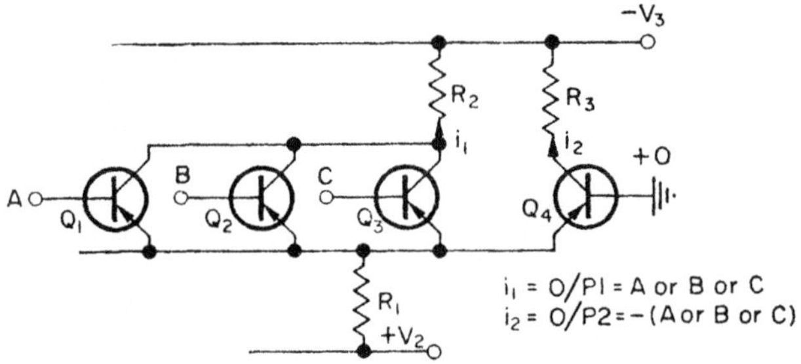

FIG. 9.17. OR and NOR circuit with three inputs.

circuitry is simple to design, with few components, the switch-over is rather slow, as the transistors are heavily bottomed when conducting.

In general any transistor will switch more quickly if it is not driven right into saturation, and this is particularly true of some of the newer high-speed transistors. To avoid saturation the collector current should be defined by a resistor in the emitter, rather than the collector, circuit. An OR circuit using this principle, with one base earthed, is given in Fig. 9.17. The common emitter current is shared between the transistors according to the base voltages. Both direct and inverted outputs are available in the form of collector currents. Resistor–capacitor coupling into a succeeding stage is not efficient, due to the large voltage translation needed.

An elegant scheme for interconnecting the all-transistor

switches uses a mixture of *npn* and *pnp* transistors, as in Fig. 9.18. This shows a three-input OR gate (*npn*) driving a three-input OR gate (*pnp*). As in Fig. 9.17, there are two outputs, one being the inverse of the other.

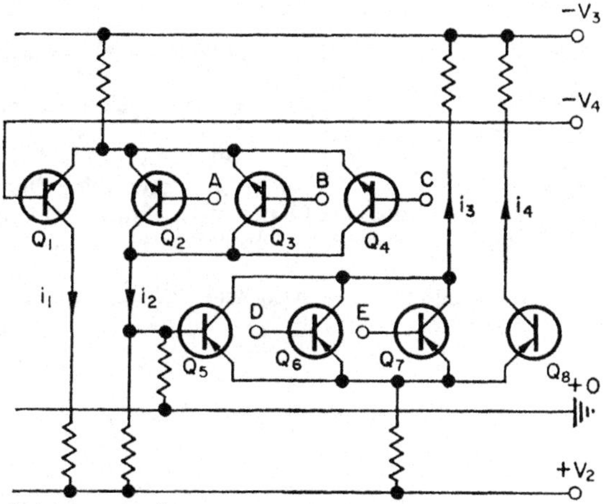

Output currents:
i_1 = −(A or B or C)
i_2 = A or B or C
i_3 = A or B or C or D or E
i_4 = −(A or B or C or D or E)

FIG. 9.18. Two-level OR circuit with complementary transistors.

Questions for Chapter 9

Q.9.1. What is meant by the binary system of numbers? Give examples of binary and corresponding decimal numbers. Explain how a bistable multi-vibrator may be used as a scale of two counter. Illustrate your explanation with a circuit diagram and explain in detail how triggering takes place.

What is the purpose of cascading bistable multivibrator circuits? How is the connection made from one circuit to the next? (I.E.E., Dec. 1959)

Q.9.2. (a) Give a careful and detailed explanation of the properties of a multivibrator.

(b) Give circuit diagrams of as many modified multivibrator circuits as possible, with a brief statement of circuit behaviour. Explain the uses of some of these circuits. (I.E.E., June 1959)

Q.9.3. Draw the circuit diagram of a typical monostable circuit and describe its operation, illustrating the answer with suitable waveform sketches.

Show how such a circuit may be used to control the duration of the forward stroke of a time-base sweep generator. (H.N.C.)

Q.9.4. Draw the circuit of a binary counting stage employing two *pnp* transistors. Outline the mode of operation of this circuit.

By means of a block schematic explain how four such stages may be joined in tandem to give a decade counter. (H.N.C.)

Q.9.5. Draw the circuit of a three-input AND gate, using diodes. Discuss in detail the requirements for successful operation of this circuit and point out how this is affected by the non-ideal properties of the semiconductor *pn* junction.

Why is an emitter-follower often used between two diode gate circuits in tandem? (H.N.D.)

CHAPTER 10

Photo-electric Applications

10.1. Types of Semiconductor Photocells

Since the commercial demand for photo-electric devices is not large, so relatively little development work has been done on these devices, in comparison with the vast effort expended on transistors and rectifiers. A few photo-sensitive semiconductor devices are available, and show greatly increased sensitivity and efficiency, when compared with the older selenium photocell. The practical semiconductor photo-electric devices can be classified as follows:

(a) Photo-conductive cells, in which an incident photon removes an electron from the valence band into the conduction band. The value of the forbidden energy gap sets a long wave limit to the sensitivity of the photocell. A photo-conductive cell appears as a linear resistance, which decreases with increased illumination.

(b) Photo-conductive diodes, operate with a reverse voltage of up to 100 V. The germanium junction photo-diode has a maximum sensitivity of about 30 mA/lumen at a light wavelength of $1{\cdot}6\ \mu$ (or 16,000 Å).

(c) Photo-transistors, have the structure of a conventional small *pnp* transistor, mounted in a suitable manner to receive incident light on the junctions. Due to the current amplification within the transistor a sensitivity of about 300 mA/lumen is possible.

(d) Photo-voltaic diodes, may be used to convert light energy directly into electric energy, without an applied voltage. The most successful of these devices is the silicon solar cell.

H*

10.2. Photo-conductive Cells and Their Applications

Germanium and silicon have too low a resistivity to be useful as photo-conductive cells, and the apparent resistance in these materials can only be raised by introducing a *pn* junction with a

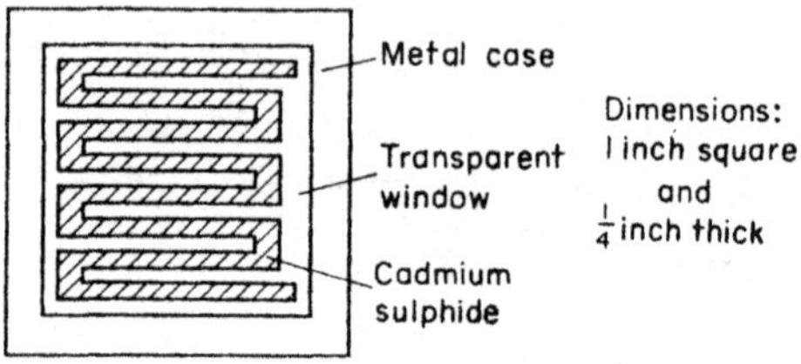

FIG. 10.1. Construction of cadmium sulphide photo-conductive cell.

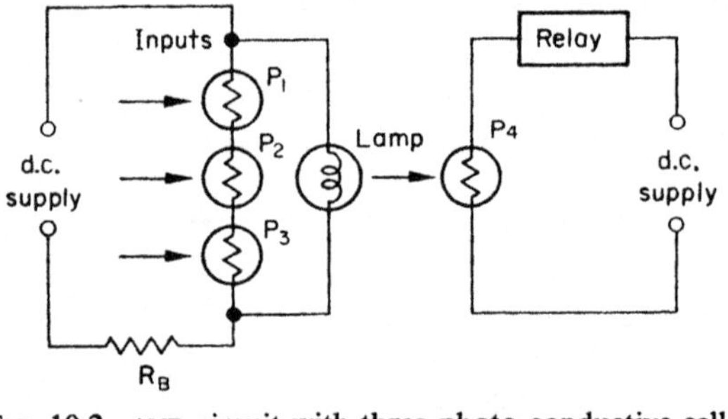

FIG. 10.2. AND circuit with three photo-conductive cells.

reverse bias voltage. Recent developments have produced photo-conductive devices with high sensitivity and considerable power rating. This enables the cell to be used to operate relays or lamps directly, without the need for an intermediate amplifier.

A typical lead selenide cell has peak response at a light wave-length of 3·3 μ (or 33,000 Å, in the infrared region), with a resistance in the range 10–100 kΩ and a maximum power dissipation of 5 mW.

A lead sulphide cell may have a peak response at 2·5 μ, with a resistance in the range 1–5 MΩ and a maximum power dissipation of 100 mW. Associated with this higher sensitivity is a time constant for the lead sulphide cell of about 50 μsec.

Cadmium sulphide cells may be obtained with a maximum sensitivity at 0·5 μ, in the visible light range, and capable of dissipating 0·5 W at a current of 50 mA. A sketch of such a cell is given in Fig. 10.1.

The lead cells have found application as infrared detectors, particularly in military equipment. Due to its relatively low cost the cadmium sulphide cell may now be considered in many industrial applications. As an example of these we may consider

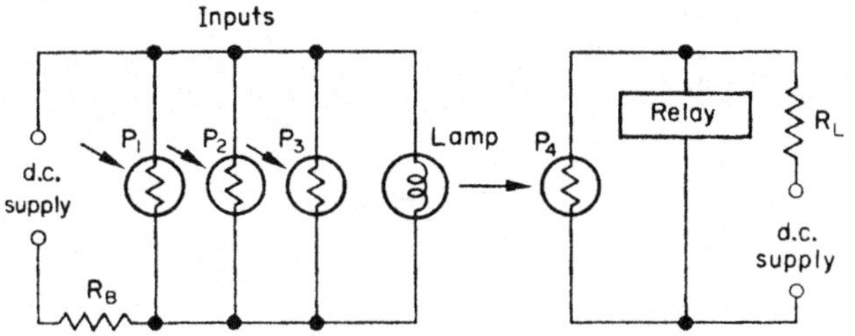

FIG. 10.3. OR circuit with three photo-conductive cells.

the AND and OR gate circuits given in Figs. 10.2 and 10.3. This AND circuit has three photo-conductive cells in series. Each cell must be illuminated simultaneously to give a low resistance path and thus extinguish the lamp. This leads to a high resistance in the output so that the series relay releases. In the OR gate a light signal on to any one of the three cells extinguishes the lamp and the cell (P_4) in the output circuit becomes high resistance. Under this condition the relay in Fig. 10.3 operates.

10.3. Photo-diode Circuit

The photo-diode may be thought of as a current source when supplied with a d.c. voltage. For large signal variations the dark current (flowing when there is no illumination) may be ignored. With small signals the dark current may be comparable with the signal current. Under these conditions it is necessary to modulate (or " chop ") the light signal, at an audio frequency, and separate

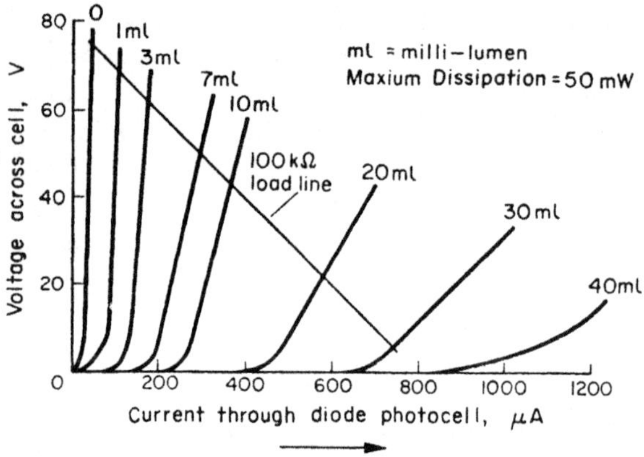

FIG. 10.4. Characteristics of a germanium photo-diode.

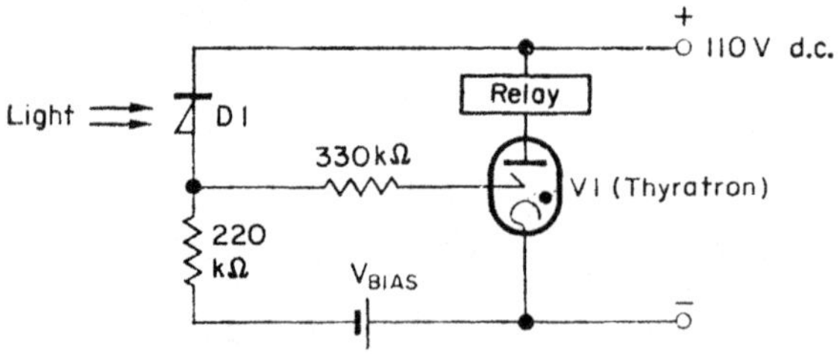

FIG. 10.5. Photo-diode circuit with thyratron and relay.

the a.c. signal and the direct dark current in the output circuit. The characteristics given in Fig. 10.4 show that a load of about 100 kΩ is suitable for a typical photo-diode.

Using a relay it is often difficult to achieve the required high impedance load, so some amplification of the photo-diode current is needed to make a useful circuit, such as Fig. 10.5. If sufficient illumination falls on the junction of the diode then the grid bias voltage is overcome, so that the thyratron can conduct. Once the thyratron conducts the relay remains operated, until the supply voltage to the thyratron is turned off.

The junction photo-diode, like the photo-transistor, has a small cross-sectional area for illumination. This is an advantage when compared with the photo-conductive cell, since the diodes may be closely packed, for applications such as the reading of punched cards.

10.4. The Photo-transistor

One of the two barriers in a photo-transistor is reverse biased and the other is biased slightly forward (both by leakage current from the collector junction and also by the external base to emitter circuit). When the base junction regions are illuminated the minority carriers formed by light photons are available to increase the emitter to collector current, in the same way as though a signal current were injected into the base.

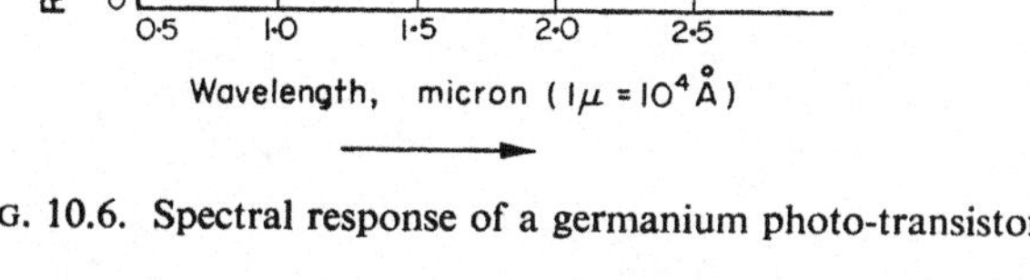

FIG. 10.6. Spectral response of a germanium photo-transistor.

The photo-transistor is thus very sensitive, due to the internal current gain effect. But the junction dark current (or leakage current) is amplified in the same way as the light signal current. Since the junction area is very small, bright illumination is needed, and this often leads to the requirement for an optical system. The germanium photocell has a spectral response to about 2 μ, which is well into the infrared region, as shown by Fig. 10.6. Unlike the typical vacuum or gas-filled photo-emissive cell, the response of the photo-transistor covers most of the frequency range of a tungsten filament lamp.

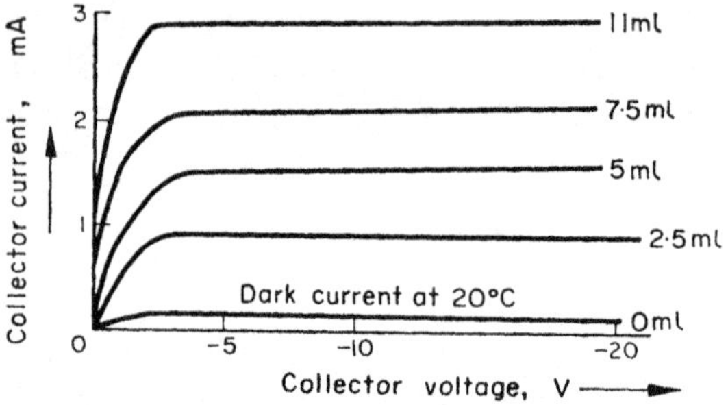

FIG. 10.7. Current/voltage curves for a germanium photo-transistor.

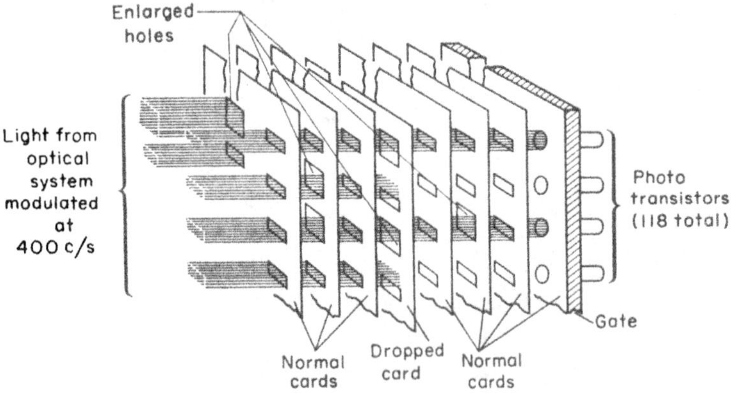

FIG. 10.8. Photo-transistors in the card translator.

The collector current/voltage characteristics of a small germanium photo-transistor are given in Fig. 10.7. This transistor has a cut-off frequency, for modulated light, of about 3 kHz. The output current, at about 12 V supply, is sufficient to operate a small relay. However, owing to the wide variations in dark current, with changes in temperature, the photo-transistor is useless for the actual measurement of light intensity, and is limited to ON-OFF applications. Such an application is illustrated in Fig. 10.8, which shows part of the card translator, used in some automatic telephone systems. This translator consists of a bin holding 1020 aluminium cards, which are mounted vertically on 40 code bars. Each card has 118 small rectangular holes, some of which are enlarged to double size to convey certain items of information.

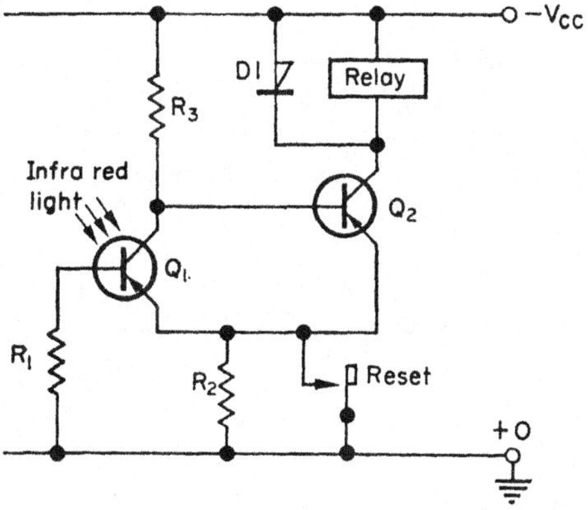

FIG. 10.9. Alarm circuit for infrared illumination.

When the appropriate code bars are operated a desired card drops a small distance. Then the modulated light beams are transmitted to the bank of photo-transistors only through the enlarged holes. The photo-transistors respond to the beams of light to operate relays, which serve to complete the telephone connection. This equipment was one of the first practical applications of transistors, and shows how large numbers may be packed into a relatively small space.

An infrared alarm circuit, using a germanium photo-transistor Q_1 and a silicon *pnp* transistor Q_2 is given in Fig. 10.9. The operation of this circuit relies on the fact that the bottoming voltage of Q_1 is less than the minimum base to emitter voltage required to bring Q_2 ON. If the reset switch is closed, short-circuiting R_2, with illumination focused on to the photo-transistor, this transistor conducts fully and holds Q_2 OFF. Immediately the infrared beam is broken the voltage across Q_1 increases, and Q_2 conducts, to operate the relay. Under this condition Q_1 cannot switch on again, since its emitter is held a few volts negative to earth, by the current through the relay and R_2.

10.5. Construction of Silicon Solar Cells

Above the atmosphere, normal to the rays of the sun, solar energy is received by a flat surface at the rate of about 1200 W/m^2. In subtropical regions, at the surface of the earth, the usable solar power is about 1 kW/m^2. The barrier layer photo-voltaic cell of

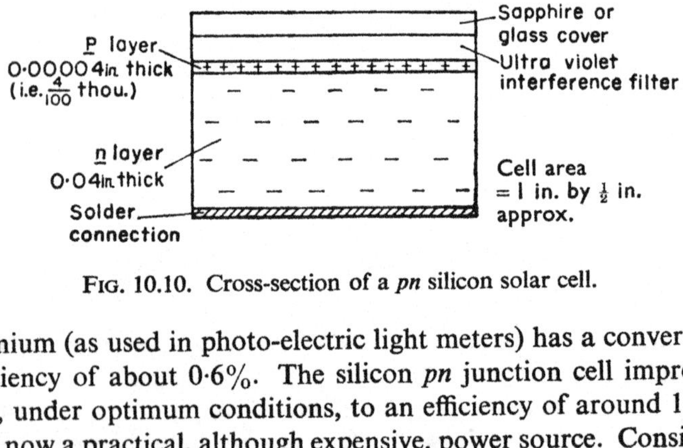

FIG. 10.10. Cross-section of a *pn* silicon solar cell.

selenium (as used in photo-electric light meters) has a conversion efficiency of about 0·6%. The silicon *pn* junction cell improves this, under optimum conditions, to an efficiency of around 12%. It is now a practical, although expensive, power source. Considerable research and development effort is being made to reduce the manufacturing cost and produce large area solar cells.

The common form of construction has a *p* layer, about 1 μ thick, on an *n* type base, with an active area of 2 cm^2. This arrangement is more susceptible to radiation damage than an *n* layer on *p*, but this latter arrangement is chemically more expensive. Radiation damage can be greatly reduced by protecting the cells with a very thin layer of sapphire. A cross-section of a typical *pn* solar cell is in Fig. 10.10. The *p* layer is diffused on to the *n* base (which consists of 1 part in 10^6 of arsenic in silicon) as a vapour of boron compounds.

The main application of solar cells is to provide power to artificial earth satellites and other space vehicles. There the cells are very rarely normal to the sun's rays, and may pass on the dark

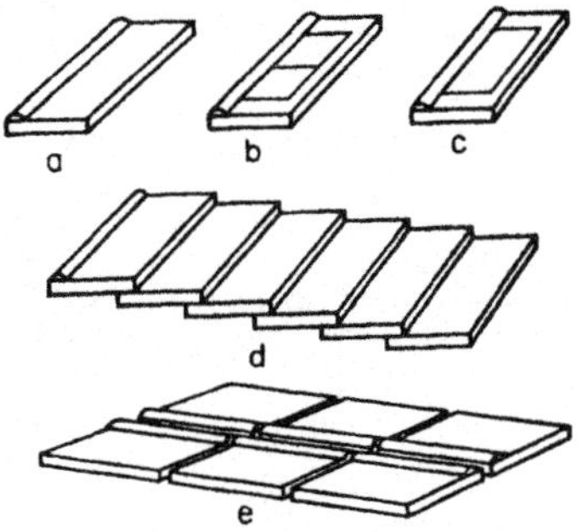

FIG. 10.11. Contacts and connections to solar cells. (a) One contact along an edge of the *p* layer. (b) and (c) Conducting grids on the *p* layer. (d) Series connection of cells. (e) Parallel connection of cells. (Reproduced with the permission of the I.E.E.)

side of the earth every orbit. The resulting 24 hr efficiency is of the order of one-tenth of the theoretical maximum power output, and the output power is usually less than 100 W.

The usual contact arrangement is to have the positive contact along one long edge of the *p* layer and the negative contact covering the whole of the underneath of the *n* layer. Cells are then joined in series in a roof-tile pattern, as shown in Fig. 10.11. Internal resistance is reduced and collection efficiency improved by having grids extending across the face of the *p* layer. Several blocks of cells are joined in parallel by bringing all the *p* contacts to a single wire and soldering the final *n* layers to a metal base.

10.6. Electrical Properties of Solar Cells

The electrical characteristics of the silicon solar cell are illustrated by Fig. 10.12. The penetration of the illuminating photons is proportional to the light wavelength. The effect of the illumination is to shift the rectifier characteristic into the fourth quadrant in this diagram. The electrical power obtainable from the cell is then the area between the rectifier curve and the axes,

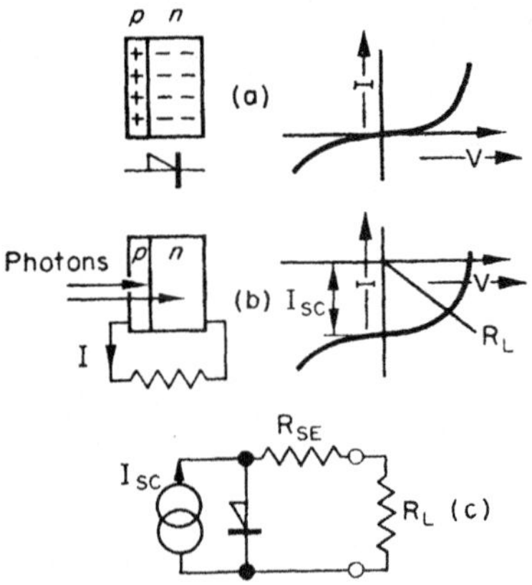

FIG. 10.12. Electrical characteristics of a solar cell. (a) In darkness. (b) Illuminated. (c) Equivalent circuit. (Reproduced with the permission of the I.E.E.)

with the load resistance R_L Ω suitably chosen to bisect this area. The current under short-circuited conditions is proportional to illumination, but the output voltage with open-circuited terminals (about 500 mV) is not very variable. The graph of Fig. 10.13 shows that at a temperature of 0°C the maximum efficiency is achieved with an output voltage of 0·5 V per cell.

The amount of energy required to produce an electron-hole pair in silicon is 1·1 eV. For a photon of light the energy is given by hc/λ, where h is Planck's constant, c is the velocity of light (3×10^8 m/sec) and λ is the light wavelength. From this we can calculate that a photon must have a wavelength of 1·1 μ, or less, to produce electron-hole pairs in silicon. The relative distribution of solar energy and frequency response for a silicon solar cell are shown graphically in Fig. 10.14.

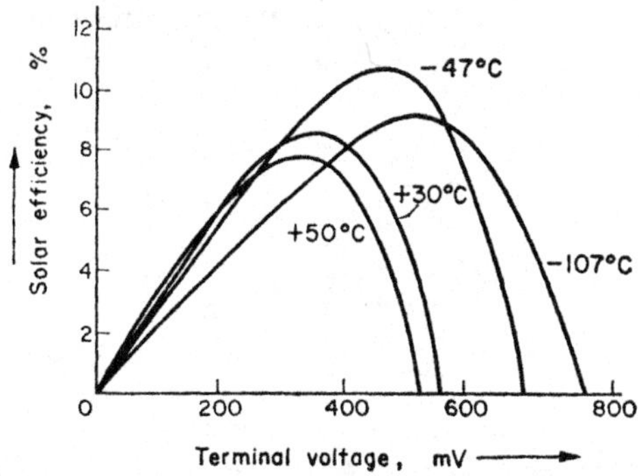

FIG. 10.13. Solar cell efficiency at various ambient temperatures and terminal voltages. (Reproduced with the permission of the I.E.E.)

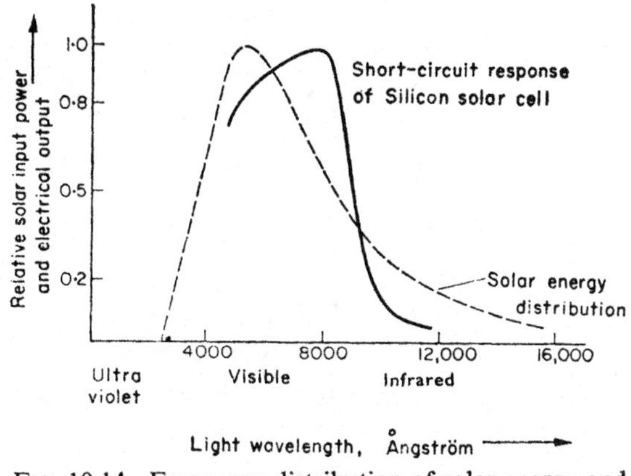

FIG. 10.14. Frequency distribution of solar energy and solar cell output power.

If a cubical space vehicle carries a group of solar cells on each of its six faces these will be connected as in Fig. 10.15. In a near-Earth orbit there will be some 6000 charge and discharge cycles each year. The series rectifiers prevent the non-illuminated cell modules from appearing as a parasitic load, and also stop any discharge of the battery through all the cells, if a satellite is in the shadow of the earth. The zener diode regulates the charging current supplied to the nickel–cadmium battery, since it conducts towards the high voltage end of the battery charging cycle.

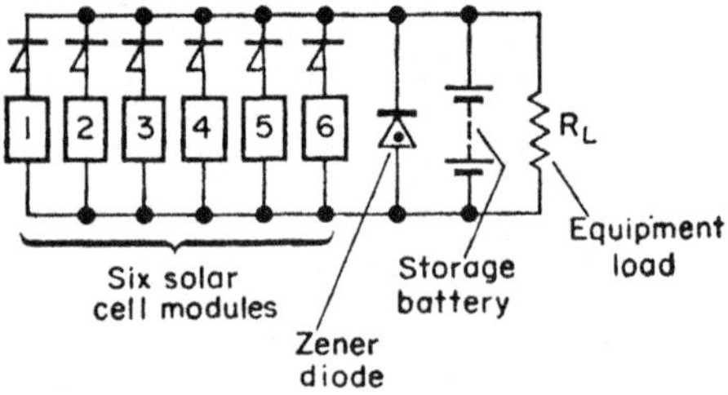

FIG. 10.15. Battery charging circuit with six groups of solar cells. (Reproduced with the permission of the I.E.E.)

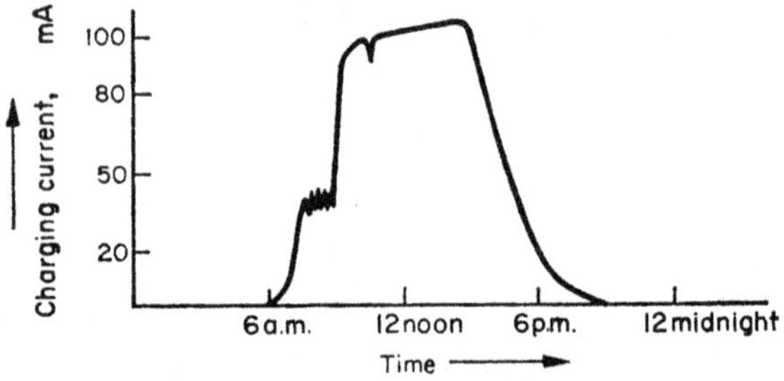

FIG. 10.16. Output from 100 cells during a September day.

For earthbound applications the solar cell has been used experimentally to power rural telephone and low power radio systems. Figure 10.16 shows the output which may be expected from an array of 100 cells, connected to give about 24 V. The graph is drawn for a sunny day in early September in the south of England, where the solar radiation is about 750 W/m^2 at noon. It can be seen that the total charge into the battery over the 24 hr period is approximately 0·5 A.hr. This is an economical source of power only under exceptional conditions.

There is some hope that further development may increase the efficiency of the silicon photo-voltaic cell to 20%. The main sources of energy loss are reflection of some of the radiant energy from the surface of the cell, the volt drop across the internal resistance of the cell, which may be reduced by epitaxial growth of the *n* layer, and recombination of electron-hole pairs within the cell.

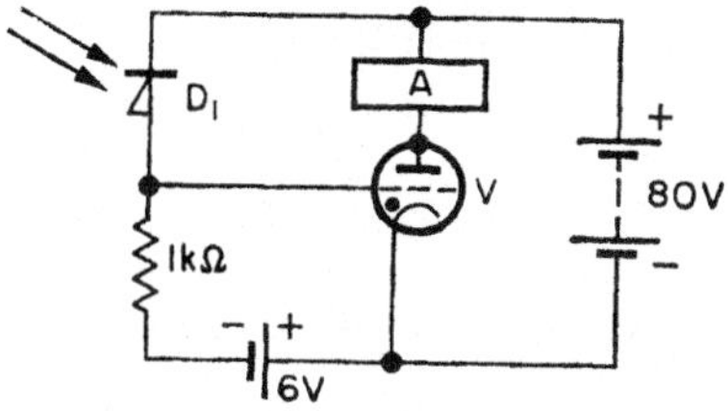

FIG. 10.17. Circuit for Q.10.2.

Questions for Chapter 10

Q.10.1. Write a short essay to distinguish between the properties of the photo-emissive cell and the photo-transistor.

Point out the limitations of each device.

Q.10.2. The photo-diode D_1 shown in the circuit of Fig. 10.17 has a constant sensitivity of 30 mA/lumen and negligible dark current. The gas-filled triode V_1 has a control ratio of 20 and with zero grid bias V_1 strikes at an anode voltage of 20 V. When V_1 conducts the anode current operates relay *A*. Estimate the minimum illumination needed at the diode junction to operate the relay, ignoring any pre-striking grid current of the thyratron.

Q.10.3. Design a bistable circuit to be controlled by a flashing light source. This circuit must incorporate two photo-conductive cells and two small neon lamps. Outline the possible mode of operation, explaining how the divide by two function, to give an electrical pulse output, is obtained.

CHAPTER 11

Special Applications

SEMICONDUCTOR devices are used in many applications which are not mentioned at all in the preceding chapters. In this chapter a few of these applications are described in outline, although no circuit is discussed in exhaustive detail. Rather, an attempt is made to emphasize the particular properties of the semiconductor device employed.

11.1. Amplitude Modulation

Modulation is the process of impressing a signal onto a carrier wave. Except for the case of the simple telephone circuit (where the carrier is direct current and the signal is voice frequency), the carrier frequency should be at least twice the frequency of the signal. If the amplitude of a sinusoidal carrier wave is modulated, then the resultant voltage is

$$v = A[1+ms(t)] \cos \omega_c t,$$

where $s(t)$ is the signal (often considered to be sinusoidal), m is the modulation index (usually expressed as a percentage) and ω_c is the angular carrier frequency (radians/second). The vector diagram and frequency spectrum for Amplitude Modulation, with a sinusoidal signal $s(t) = \cos \omega_s t$, are shown as (a) and (b) in Fig. 11.1. The A.M. waveform, showing how the carrier " envelope " varies with the modulating signal, is sketched as part (c) of that same figure. When the modulating signal consists of a band of frequencies, then each component frequency produces two side frequencies, as shown in Fig. 11.1(b), and these form the upper and lower sidebands.

There are a number of possible amplitude modulator circuits using transistors or diodes as the active element. In each case the amplitude modulation is produced by a non-linear characteristic of the device as it is used in the circuit. A simple modulator, with a single transistor, is shown in Fig. 11.2, where the transistor Q_1 may be a general purpose germanium device for frequencies up to about 100 kHz. With the component values as shown the circuit is adjusted for a carrier frequency of 20 kHz and a passband of

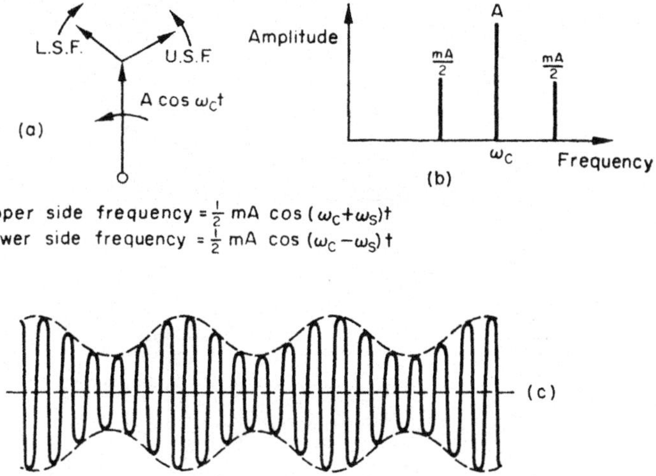

FIG. 11.1. Vector diagram, frequency spectrum and waveform for amplitude modulation.

slightly more than ± 2 kHz. This is suitable for line transmission and also ideal for demonstrating the effects of amplitude modulation with a large screen oscilloscope.

Amplitude modulation (together with unwanted second harmonic distortion) is produced by the circuit of Fig. 11.2 since the transistor has a collector current/base voltage characteristic which contains a square law term. We will assume that the base to emitter d.c. voltage is such that

$$i_c = 40\, v_{be} + 2(v_{be})^2 \text{ milliamps,}$$

where v_{be} is measured in volts and consists of a carrier of 2 V peak at 20 kHz and a sinusoidal signal of 1 V peak at 1 kHz. That is,

$$v_{be} = 2 \sin 2\pi(20 \times 10^3)t + 1 \sin 2\pi(10^3)t \text{ volts.}$$

Then the resulting collector current is,

$$i_c(\text{mA}) = 80 \sin 2\pi(20 \times 10^3)t + 40 \sin 2\pi(10^3)t + 2[2 \sin 2\pi(20 \times 10^3)t + 1 \sin 2\pi(10^3)t]^2.$$

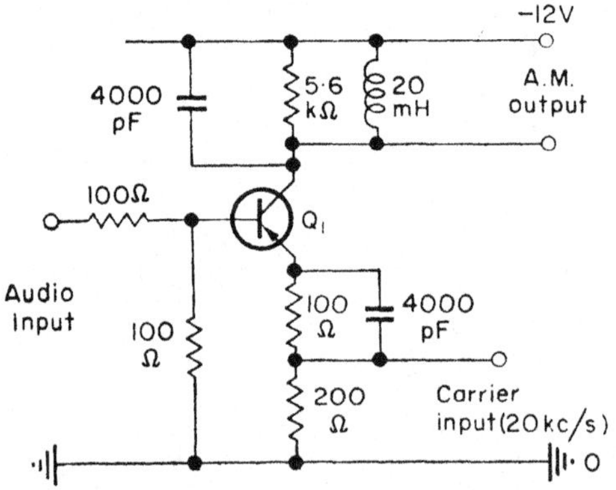

FIG. 11.2. Simple amplitude modulator circuit.

The first two terms represent straightforward amplification, and due to the presence of the tuned circuit, the 1 kHz term will not produce a significant output voltage. The collector current expression can be rewritten as

$$i_c\,(\text{mA}) = 80 \sin 2\pi(20 \times 10^3)t + 40 \sin 2\pi(10^3)t + 8 \sin^2 2\pi(20 \times 10^3)t + 2 \sin^2 2\pi(10^3)t + 8 \sin 2\pi(20 \times 10^3)t \sin 2\pi(10^3)t.$$

The two terms involving the square of a sine wave can each be resolved into a d.c. term and a second harmonic term. None of these will have any significance in the output. The first and last terms represent the carrier wave and the two side frequencies, respectively, and can be expressed as the A.M. wave,

$$i_{A.M.} = 80[1+0{\cdot}1 \sin 2\pi(10^3)t] \sin 2\pi(20\times 10^3)t.$$

Hence the output consists of a carrier wave and two side frequencies, at 19 kHz and 21 kHz, with a modulation index of 0·1, or 10%. This is known as double sideband amplitude modulation, with transmitted carrier.

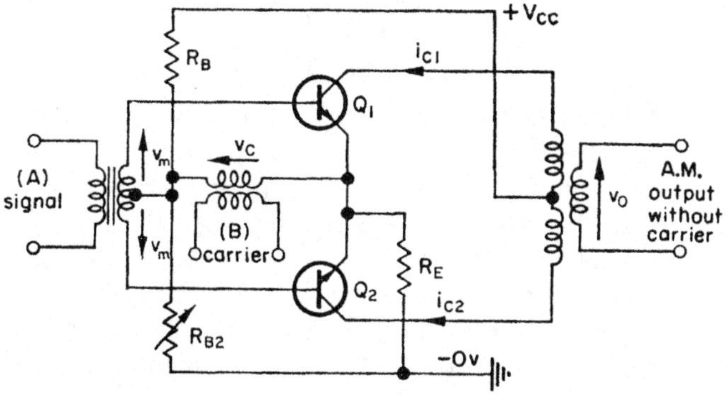

FIG. 11.3. Balanced transistor modulator to suppress carrier.

In fact the carrier frequency carries no useful signal information, although it represents a large proportion of the total power in the A.M. output. To conserve transmitter power it is a great advantage either to reduce or to eliminate the carrier, and this is achieved by the suppressed carrier circuit of Fig. 11.3, which actually employs a matched pair of *npn* transistors (although a circuit with *pnp* transistors would do equally well). In this circuit the modulating signal is applied at input *A* and the carrier frequency at input *B*, to produce the voltages v_m and v_c, as shown. Due to the centre-tapped input transformer the bases of the two transistors will receive the voltages,

$$v_{b1} = v_c + v_m$$

and $$v_{b2} = v_c - v_m, \text{ at any instant.}$$

If the base bias and the input voltages are adjusted such that the transistors are operating on the square law portion of the irput characteristic, then,

$$i_{c1} = a+b(v_c+v_m)+c(v_c+v_m)^2$$

and

$$i_{c2} = a+b(v_c-v_m)+c(v_c-v_m)^2,$$

where a, b and c are constants of the two transistors. As in all push-pull amplifiers, the output voltage is proportional to the instantaneous difference between collector currents,

or $$v_0 = K(i_{c1}-i_{c2}).$$

Hence $$v_0 = K(2bv_m+4Cv_mv_c).$$

The modulation frequency term $(2bv_m)$ is well removed in frequency from the carrier, and can be eliminated by a simple tuned circuit. The second term $(4Cv_mv_c)$ produces the two sidebands, equally spaced in frequency on either side of the carrier. It is common practice to transmit only one sideband, thus giving a Single Sideband, or S.S.B., system. In this case the unwanted sideband may be removed by a band-pass filter circuit in the output from this circuit (often known as a balanced modulator).

Diode modulator circuits are often employed in carrier telephone systems, where large numbers of modulators are required, the output power to the line is of the order of milliwatts and low cost is essential. Due to the rectifier action of the diodes there is considerable harmonic content in the output from the modulator, but this may be eliminated by correct choice of carrier frequency and the use of appropriate band-pass filter circuits. For instance, if twelve carrier frequencies between 60 kHz and 108 kHz are used, with a maximum audio modulating frequency of 3·4 kHz, then the harmonics of carrier and audio signal frequencies are well outside the transmitted band. Two simple diode modulator circuits are shown as (a) and (b) in Fig. 11.4. For correct operation of these circuits it is essential that the diodes be matched in both forward and reverse characteristics. Copper oxide rectifiers (or " varistors ") are still widely used for telephone modulator application, since the cost is low and the resistance characteristic is very suitable. The applied carrier voltage must be larger than the maximum signal voltage, so that the diodes are biased on and off by the carrier and not by the signal. In modulator (a) the carrier voltage causes the diode bridge to present a low or very high resistance, in series, to the signal source in alternate half-cycles of carrier frequency. In circuit (b) there is an alternating effective shunt low or high resistance at the carrier frequency. The result, in either case, is to transmit the audio signal for half the period of the carrier wave, as sketched in part (c) of Fig. 11.4.

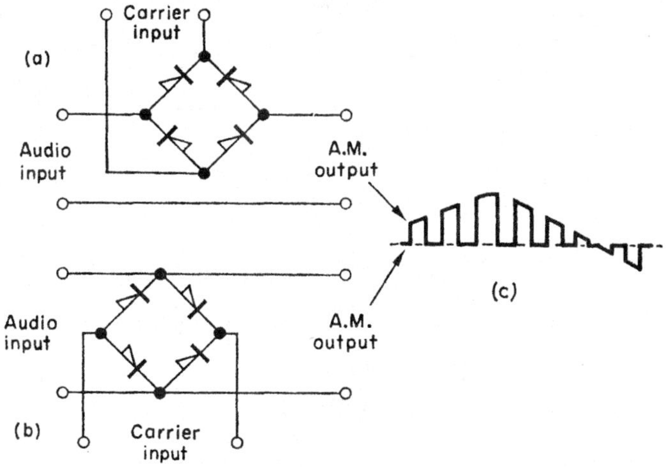

FIG. 11.4. Diode modulator circuits with output waveform.

The output has effective audio and carrier suppression with almost half the signal voltage being transmitted. The sharp transitions in the output waveform represent the presence of harmonic frequencies, which are removed by filtering.

11.2. Amplitude Demodulation
(Detection and Frequency Changing)

Detection is essentially rectification of the amplitude modulated signal, to produce a voltage varying in amplitude in accordance with the envelope of the modulated carrier. Diode circuits are most important for this operation. The simple diode detector circuit is given in Fig. 11.5 and the demodulating action is illustrated by the sketch of Fig. 11.6. Correct operation of this circuit requires a load circuit with a time constant that is large

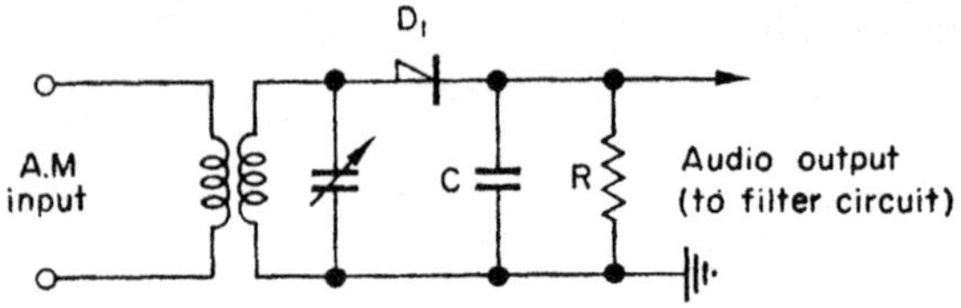

FIG. 11.5. Diode detector circuit.

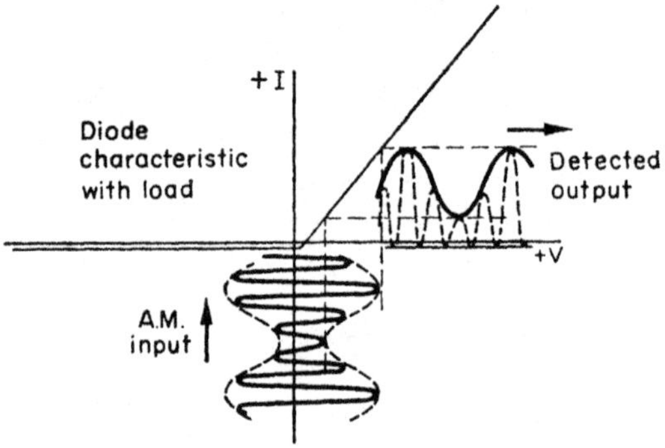

FIG. 11.6. Diode characteristic with A.M. input and audio output.

compared with the period of the carrier and yet small compared with the signal frequency. That is,

$$\omega_c \gg \frac{2\pi}{CR} \gg \omega_s.$$

In a typical medium frequency broadcast receiver the detector is followed by a simple resistance–capacitance R.F. filter network and a high resistance potentiometer, to act as a volume control.

A linear detector has an output voltage which is directly proportional to the amplitude of the envelope of the input wave.

The detection efficiency of a diode detector can be measured directly at mains frequency, with the circuit of Fig. 11.7. The capacitance C must now be large enough to act as a by-pass at mains frequency, rather than at the carrier frequency for which the diode detector is normally used. The diode load resistance R is kept at the normal circuit value. The r.m.s. input voltage v_1 is measured by a high impedance voltmeter and the direct current I_0 through the load is indicated by a microammeter. If the input voltage is varied by a potentiometer, then a graph of I_0 microamps for v_1 volts can be plotted, to give the ratio I_0/v_1. The output

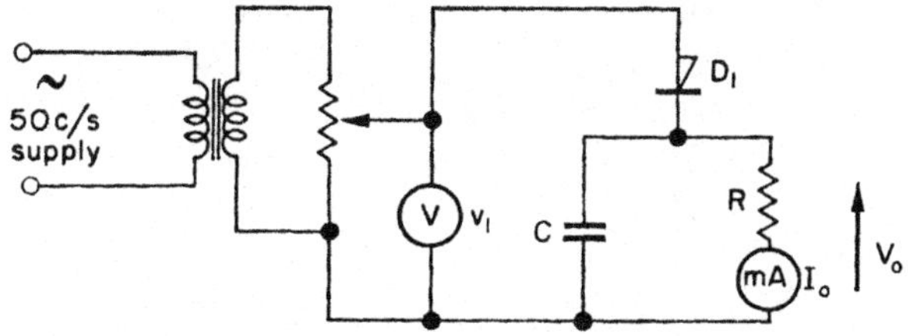

FIG. 11.7. Circuit to measure detection efficiency.

voltage for the detector is $V_0 = RI_0$ and the peak input modulated amplitude is $\hat{v}_1 = v_1\sqrt{2}$. Then the detection efficiency is given by

$$\eta_{\text{det.}} = \frac{V_0}{\hat{v}_1} = \left(\frac{I_0}{v_1}\right)\left(\frac{R}{\sqrt{2}}\right).$$

If the incoming A.M. wave is a single sideband, then the carrier needs to be reintroduced locally at a balanced type of detector to give the correct output. To avoid frequency distortion of the signal, the locally generated carrier must be very close in frequency to the original carrier. The modulating and demodulating circuits for single sideband systems are almost identical. For this reason a circuit package, containing a bridge of four suitable diodes, is sometimes known as a " modem ".

Frequency changing may be regarded as another form of demodulation, in which signals at two frequencies, often widely removed, are mixed in a non-linear device to give an output at the difference frequency. This principle is used in almost all A.M. systems, whether for line or radio transmission. At microwave frequencies, in both radio communication and radar systems, frequency changing is particularly valuable, since amplification

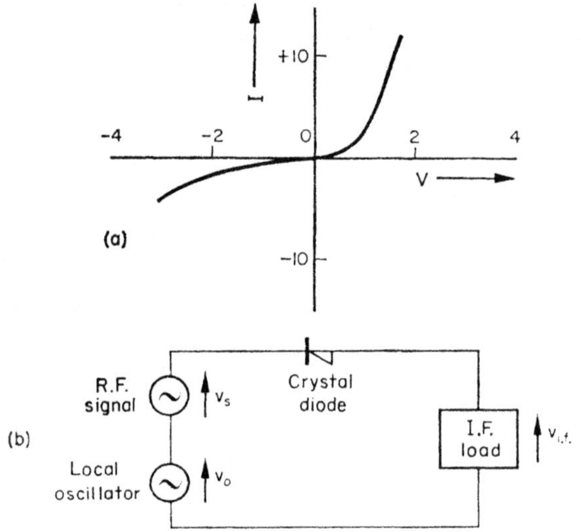

FIG. 11.8. Microwave crystal frequency changer. (a) Current–voltage characteristic. (b) Basic circuit.

with low noise is difficult at U.H.F. and S.H.F. Silicon crystal point-contact diodes are very useful at these frequencies, as there is practically no transit time effect and the noise figure, although it may approach 10 db at 10 GHz, is usually lower than that of available microwave amplifiers, at reasonable cost. Only small amplitude signals can be accepted, but the signal level at the input to a microwave receiver is usually very low. The current/voltage characteristic for a microwave silicon crystal and the basic frequency changing circuit are shown as (a) and (b) in Fig. 11.8.

The amplitude of the output from the local microwave oscillator is large with respect to the received signal, and the impedance Z_{IF} represents the input impedance of the first Intermediate Frequency amplifier. A typical I.F. is 70 MHz, at which frequency wideband transistor amplifiers are available. The silicon point-contact mixer diode is usually operated without a d.c. bias voltage. In certain microwave applications a germanium junction diode gives a lower noise figure, but the device capacitance is larger and a small bias voltage is needed.

11.3. Frequency Modulation

With frequency modulation the instantaneous angular velocity of the carrier wave is dependent on the instantaneous amplitude of the signal wave. Therefore the resulting frequency varies on either side of the unmodulated frequency, at a rate dependent on the frequency of the modulating signal. For a sinusoidal modulating signal the instantaneous angular velocity of the F.M. wave is given by

$$\omega_i = (\omega_c - \Delta\omega \cos \omega_s t) \text{ rad/sec},$$

where ω_c = angular velocity of the unmodulated carrier wave,

ω_s = angular velocity of the modulating signal, and

$\Delta\omega$ = maximum deviation of ω_i from ω_c.

The function of the frequency modulating circuit is to ensure that the deviation $\Delta\omega$ has a linear relationship to the signal amplitude.

The ratio of deviation to signal frequency is known as the Modulation Factor, or

$$m_f = \frac{\Delta\omega}{\omega_s} = \frac{\Delta f}{f_s}.$$

So it can be seen that, for a given deviation, the modulation factor varies inversely with the modulating frequency. Frequency modulation gives rise to a number of sideband frequencies, spaced symmetrically on either side of the unmodulated carrier frequency, where the number of significant sidebands is dependent on the modulation factor m_f. In a typical F.M. broadcasting system the maximum audio signal frequency is 15 kHz and the deviation is 75 kHz, giving a modulation factor of $m_f = 5$.

Two different methods of producing F.M. are in common use. In the more direct method the reactance of a tuned circuit is made to vary with the amplitude of the applied signal, and thus the output frequency of an oscillator is modulated. An example of this is the reactance valve circuit, which commonly employs a

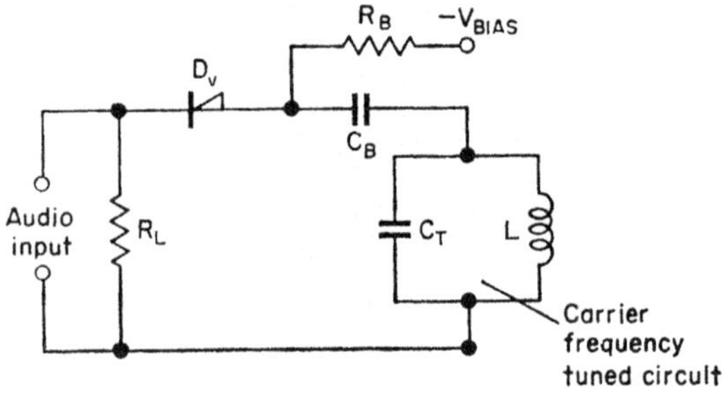

FIG. 11.9. Frequency modulator circuit with a varactor.

pentode having a " variable μ " characteristic. The use of a transistor in this type of circuit has not proved very successful due to two factors. Firstly, it is difficult to find a transistor with a suitable characteristic to give a gain varying with input voltage in a linear manner. Secondly, due to the small permissible swing of input voltage at the base of the transistor, the resulting deviation is too small.

The variable capacitance diode (or " varactor ") is suitable for use in a reactance modulator associated with a low power F.M. transmitter. This is illustrated by the simple circuit of Fig. 11.9, where the varactor is reverse biased by the d.c. supply voltage

V_{CC}. A table showing how the measured junction capacitance of a typical varactor varies with reverse bias voltage is given as (a) in

(a)

Reverse bias $-V_{BIAS}$, V	Diode capacitance C_D, pF
0·5	315
1·0	270
2·0	225
3·0	199
4·0	182
5·0	169
6·0	160
7·0	151·5
8·0	145
9·0	140
10·0	135
11·0	130
12·0	124·5
13·0	120
14·0	116

(b)

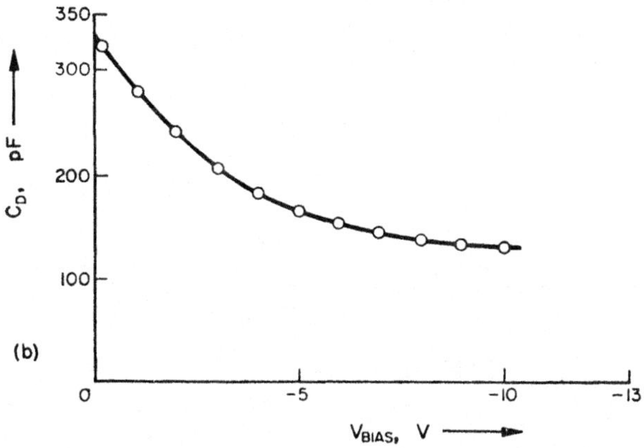

FIG. 11.10. Measured junction capacitance of a varactor.

Fig. 11.10 and the same results are plotted as a graph in (b) of Fig. 11.10. It is clear from this graph that, although a wide range of capacitance is available, the capacitance/applied voltage relationship is not a straight line. However, since the resonant frequency of a tuned circuit varies in proportion to the square root of capacitance, it is possible to obtain narrow deviation frequency modulation within reasonable distortion limits. The properties and applications of varactors are considered in more detail by Registrar (see Appendix B).

Frequency modulation can also be produced first by phase modulating the carrier wave with the signal voltage and then converting the phase modulation to frequency modulation. With phase modulation the phase angle of a carrier wave is advanced

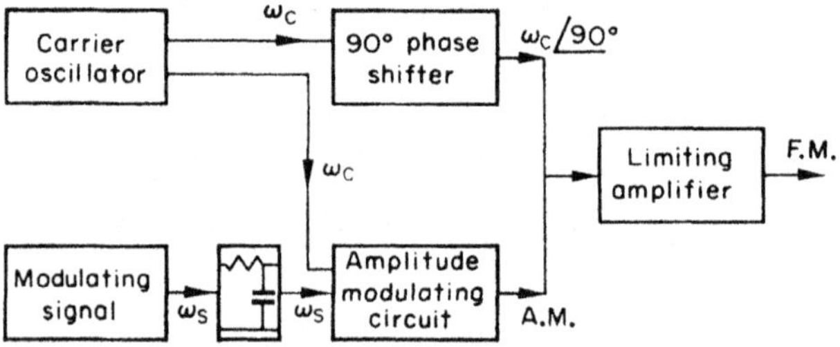

FIG. 11.11. Simple phase modulator.

or retarded at a rate equal to the frequency of the modulating signal. This phase shift is proportional to the amplitude of the modulating signal. It follows that for Phase Modulation the frequency deviation is given by the equation

$$\Delta_f = \text{phase swing} \times \text{modulating frequency.}$$

The peak deviation is, therefore, only obtained at the highest modulation frequency and, for a given modulation amplitude, the deviation is lower at the lower modulating frequencies. So phase modulation may be converted to frequency modulation by inserting a network, with an attenuation that increases with frequency, into the modulating signal section of the circuit before the phase modulator. A simple phase modulator is shown in Fig. 11.11. Here an amplitude modulated wave, produced by a transistor modulator circuit, is combined with a somewhat larger unmodulated carrier wave, which is shifted in phase by 90° from the phase of the carrier in the A.M. wave. The phase of the resulting voltage vector then varies in accordance with the amplitude of the modulating signal and any residual amplitude modulation is removed by a limiting amplifier. This simple circuit gives linear modulation to a phase shift of about $\pm 15°$. The Armstrong Phase Modulator circuit employs a balanced amplitude modulator (using four diodes) in the arrangement of Fig. 11.11, and linear

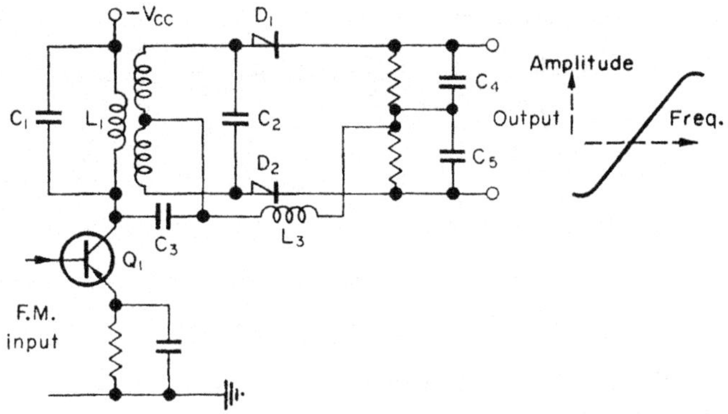

FIG. 11.12. Frequency modulation discriminator.

modulation is then available to a phase shift of $\pm 30°$. Since the amplitude modulation may be carried out at a relatively low carrier frequency, the phase modulation process of producing frequency modulation is more satisfactory with semiconductor devices than the direct F.M. method. The phase shift produced is very small when compared with the frequency deviation required in practice, so a large degree of frequency multiplication is required after the phase modulation stage.

Two slightly different circuits, both requiring a pair of matched diodes, are in common use for the detection of frequency modulated signals. The F.M. discriminator circuit is illustrated in Fig. 11.12. This circuit commonly operates with a carrier frequency at H.F. or V.H.F. and this imposes certain requirements on the transistor and diodes which are employed. When the discriminator is correctly adjusted, an S-shaped curve, of output voltage for input frequency, is obtained, as the incoming frequency is varied equally on either side of the nominal centre frequency. The output of the discriminator is sensitive to the amplitude of the input and so the input should be derived from a limiting amplifier, to remove the possibility of this type of distortion.

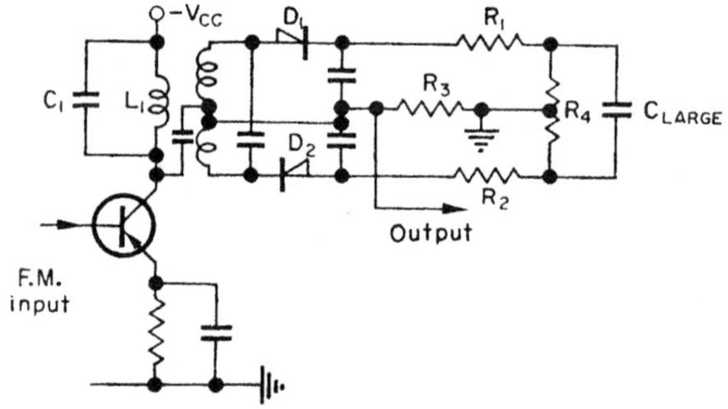

FIG. 11.13. Ratio detector circuit.

The ratio detector differs from the discriminator only in that one diode is reversed in polarity and the output is taken between earth and the centre-tap of a high value resistor R_2 as shown in Fig. 11.13. This circuit can be designed, mainly by correct tapping adjustment of the transformer, to be unresponsive to amplitude modulation, whilst giving a similar S-shaped detection curve, with a frequency modulated input, as the discriminator. However, the output voltage (per volt of input signal) is only half as great as with the discriminator. Since small signals are involved, and a low device capacitance is essential, small germanium point-contact diodes are ideal for these two F.M. detector circuits.

11.4. Electronic Switching Circuits

This section is concerned with electronic switching, with particular reference to the applications in electronic telephone exchanges. In a telephone switching system, arrangements are made to interconnect any two telephones to give a speaking path.

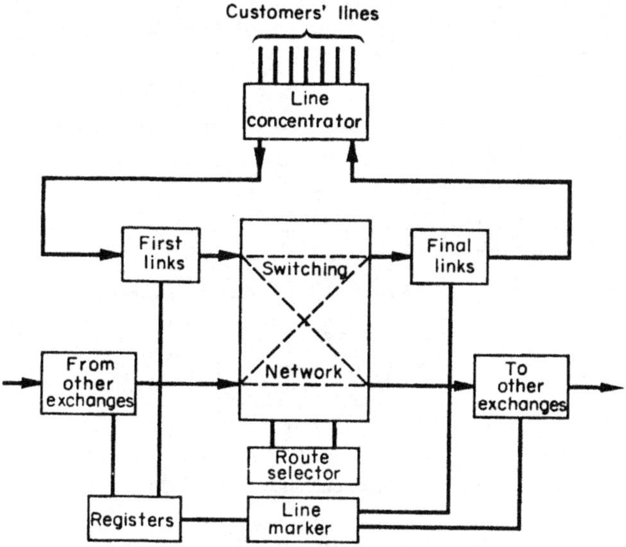

FIG. 11.14. Schematic of a space division telephone exchange.

This requires a large number of " crosspoints ", where the input and output paths may be interconnected. The present trend of development would appear to favour the " Space Division " type of electronic telephone exchange in which the crosspoints carry only audio frequencies. Existing telephone exchanges use electro-mechanical relays as the crosspoints, with four relay contacts usually required at each crosspoint. Higher speed operations, which are expected to lead to a lower overall cost, can be given with electronic devices as the crosspoints. Gas diodes or triodes are being used in small exchanges, but have the disadvantages of large volt drop, relatively large power consumption and considerable volume. *pnpn* diodes or transistors show advantages in these respects, but are relatively expensive in comparison. Semiconductor devices may be used in all parts of an electronic telephone exchange, for which a block schematic is given in Fig. 11.14, but the majority of the devices will be in the central Switching Network and this is certainly the most decisive in regard to the cost of the system.

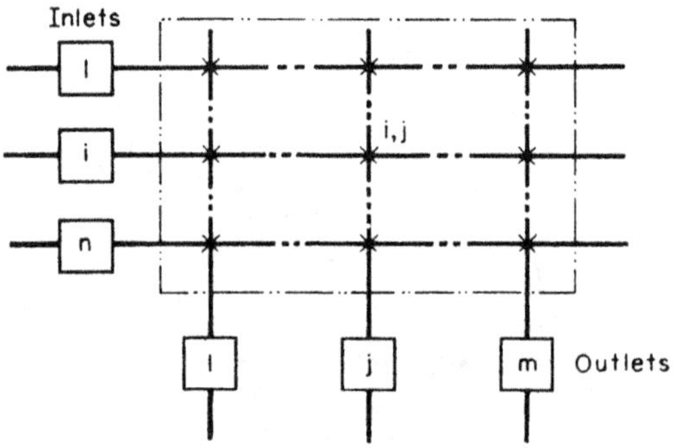

FIG. 11.15. Typical crosspoint switching array.

A typical switching array is one in which n inlets have access to m outlets, where n is often a larger number than m. As indicated by Fig. 11.15 this gives a requirement for $m \times n$ crosspoints. To have a connection between inlet i and outlet j it is necessary to operate the crosspoint ij. While this crosspoint is in use, no other points on the i or j coordinates may be operated. For our purposes, with semiconductor devices, a crosspoint is operated when it is switched from a high impedance to a low impedance state. The path through the crosspoint must conduct at least the audio frequency band (say 200–4000 Hz), with low electrical noise, and must be simple to operate and hold operated.

The current/voltage graph for a small *pnpn* diode is given as Fig. 11.16. The *pnpn* diode may be triggered ON by applying a short voltage pulse to the transmission path. Its maintaining voltage V_M is quite low, and is typically of the order of 1 V. The striking voltage is V_S, and the voltage V_B is the battery voltage supplied to the circuit. Figure 11.17 shows a single switching stage using such a *pnpn* diode. The two triggering pulses, of $-V_P$ and $+V_P$, add so that $2V_P = 1{\cdot}5\ V_S$ to ensure that the diode

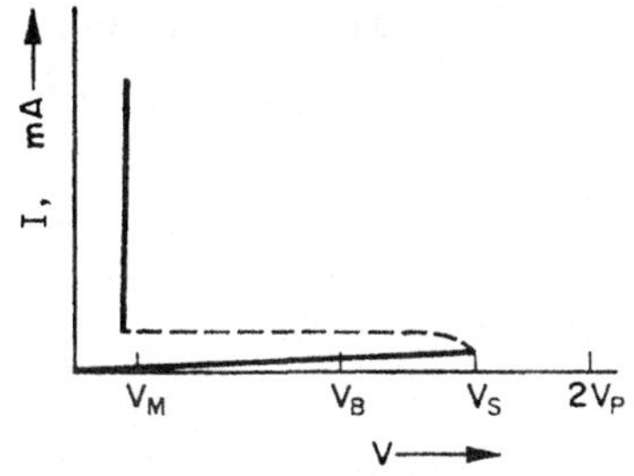

FIG. 11.16. Characteristic of a *pnpn* diode.

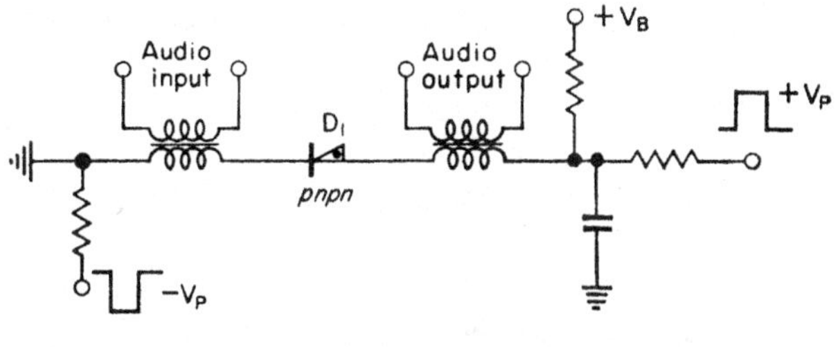

FIG. 11.17. Single switching stage with one *pnpn* diode.

conducts. The germanium *pnpn* diode does not have a very high impedance when OFF and this may lead to excessive crosstalk between speech channels. The silicon *pnpn* diode is better in this respect. For both types of small *pnpn* diode the power handling capabilities are limited, and it is just possible that a high speed voltage may cause a false breakdown to the conducting state. The switching time of a small *pnpn* diode is of the order of 2 μsec, which is faster than the rate at which individual calls arrive at a typical 10,000 line exchange. This leads to the important principle of " one-at-a-time " switching, and is a significant improvement over the slower gas-filled tubes. The requirement for input and output speech transformers is an expensive part of this type of switching.

The four-layer *pnpn* transistor has some advantages as a crosspoint over the diode, to be set against the higher initial cost. The transistor may be triggered externally from the speech path, by a small pulse to the base lead, and is then held ON by a relatively

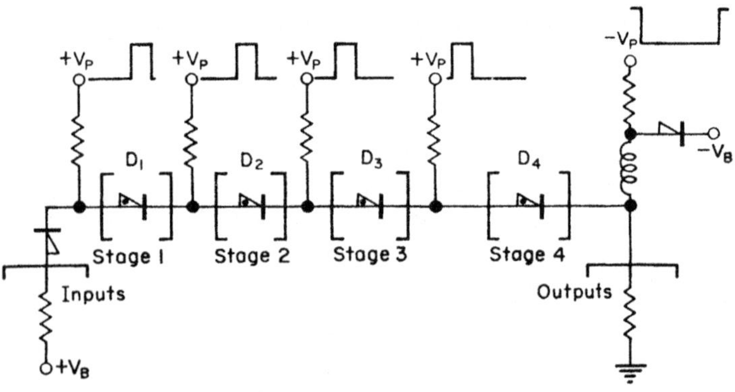

FIG. 11.18. Four-stage switching circuit with many possible paths.

low supply voltage. Also the transistor gives gain in the forward direction, whereas the diode is approximately a 20 Ω resistance when ON. Otherwise, the circuit arrangement is much the same for either diodes or transistors.

To provide for a large total number of inputs and outputs, with the minimum number of crosspoints, it is usual to do the telephone switching in a number of sequential stages, typically with a 10 : 1 switch ratio in each stage. So a 10,000-line exchange then needs four switching stages. The pulses to control these stages are derived from a central Register and Route Controller. As the switching time through the exchange using semiconductor devices is so small, calls are set up on a one at a time basis. The possible sequence is indicated by the diagram of Fig. 11.18. The square brackets (E) in that diagram show the points where a number of possible paths are joined in common, at the input and outputs to each stage. It is easy to visualize a system where each switching stage has one hundred inputs and ten outputs, so that ten calls can be in progress through each stage simultaneously.

A specified path is triggered into conduction by first applying a negative pulse to the output of that path. Positive pulses are then applied sequentially to the chosen diodes, working backward at 2 μsec intervals from the output. So diodes D_4, D_3, D_2 and D_1 break down in turn, each to about 20 Ω forward resistance. The calling and called telephone lines are then joined by a low resistance path, within 10 μsec after the complete number has been dialled.

11.5. Ferrite Core Driving Circuits

Ferrite magnetic rings, or cores, are used as information storage devices in many digital computers, and at the present time are the most versatile and reliable form of electronic memory which is available. The ferrite core can be designed to have a rectangular magnetic hysteresis loop, such as that in Fig. 11.19. For use in the digital computer system one state of magnetization is designated " 1 " and the other state " 0 ", and the storage properties depend on maintaining a nearly rectangular hysteresis loop. A current of

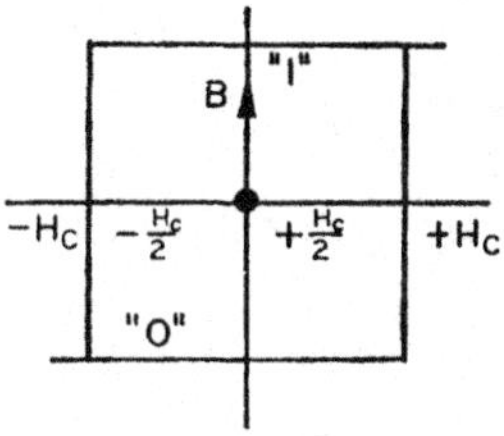

FIG. 11.19. Rectangular hysteresis loop of a ferrite core.

sufficient magnitude, corresponding to $+H_C$ in one direction through a winding on the core, will leave the core magnetized in the state representing " 1 ". A current through the core winding in the opposite direction will, if sufficiently large, reverse the direction of the flux and the magnetization of the core will then correspond to " 0 ".

A common size of moulded ferrite core is 60 mil outside diameter, 45 mil inner diameter and 20 mil axial length (where

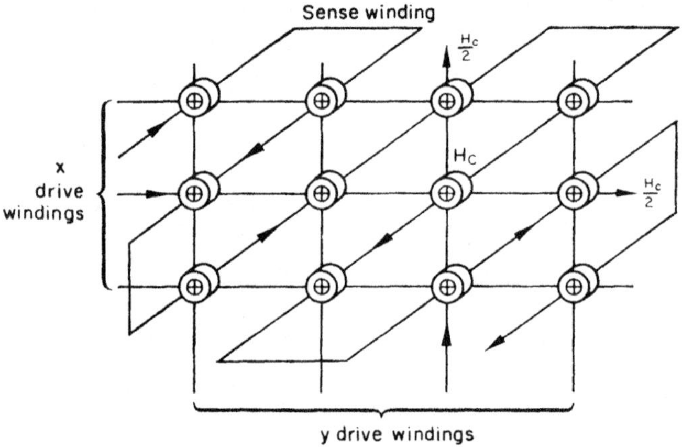

FIG. 11.20. Part of a ferrite core matrix array.

1 mil = 0·001 in.). Smaller cores, down to 20 mil outside diameter, are available. However, the smaller cores are fragile, and the hysteresis loop is less rectangular, as the ratio of inner to outer diameter is decreased.

The cores are usually arranged in a symmetrical matrix array, typically numbering 64 × 64, or 128 × 128. Each core in the array is specifically defined by two co-ordinate wires threaded through the cores at right angles. Figure 11.20 shows a small part of such a ferrite core matrix. To detect the presence of a 1 or 0 in any particular core a " sense winding " is threaded through the cores in series.

The ferrite core of 60 mil diameter requires nearly 1 ampere-turn to be switched from one state to the other. Since a wire through the centre of the core represents just one turn, so comparatively

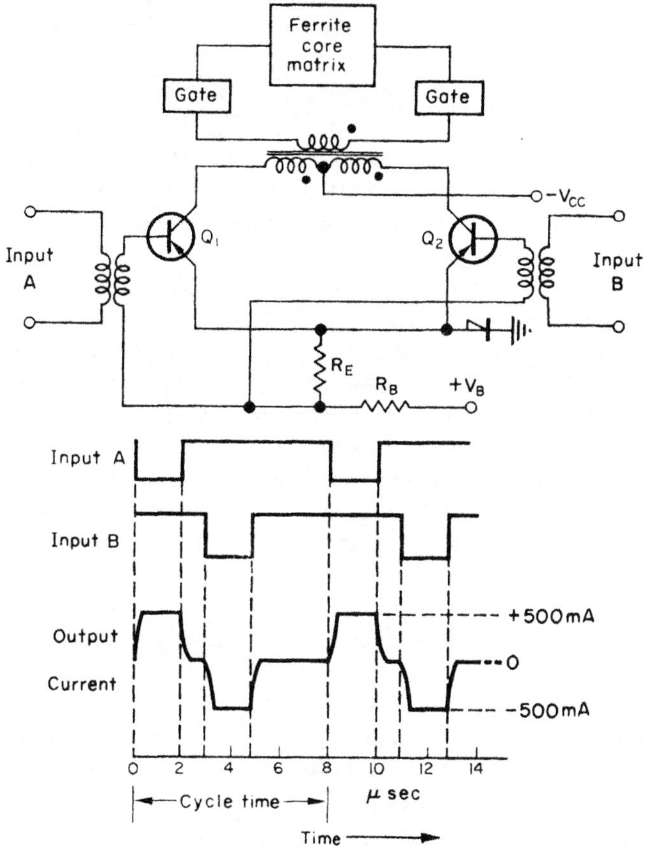

FIG. 11.21. Pulse amplifier to drive ferrite cores.

large currents are required to switch states. " Writing " and " Reading " are both done by sending current pulses on the two wires (*X* and *Y*) which coincide at the required core. Other cores on these two wires are then only " half-switched " and do not experience a change of state. With a typical switching time for the core of the order of 2 μsec and the need for a current pulse approaching 0·5 A, this imposes fairly difficult conditions on the design of suitable amplifiers using transistors. The rise time of the driving pulse must be a fraction of a microsecond and the turn off time for the pulse must not be greater than 1 μsec to give a short cycle time, so that pulses to one core do not interfere with following pulses. In some small stores high-speed ferrite cores have a switching time of 0·5 μsec and a complete cycle time of 2 μsec is then possible. A suitable form of driving circuit is given in Fig. 11.21, together with the input and output waveforms. The input pulses, at driving points *A* and *B*, are derived from transistor multivibrator circuits. The input transformers have a suitable turns ratio to give a driving pulse of about −5 V at the bases of two transistors. The output transformer consists of three identical windings on a ferrite core and the gate circuits control the application of the output current pulses to the *X* and *Y* drive windings as appropriate. This diagram shows a cycle time of 8 μsec, a pulse length of nearly 2 μsec and a pulse amplitude of 500 mA. The use of coincident current driving gives economy in drive circuits, since only 128 driving amplifiers are required to switch any one of the 4096 ferrite cores in a 64 × 64 matrix.

To find whether a 1 or 0 is stored in any core the coincident *X* and *Y* currents are sent through that core to set it to 0. If the core is already in the 0 state, then there will be no change of state and no output on the sense winding. But if the core was magnetized in the 1 state, then the flux reverses and an output of about 50 mV is obtained on the sense winding. At the same time the act of sensing destroys the one " bit " of information stored in the core.

11.6. D.C. Motor Control

Battery driven d.c. motors are used for a number of purposes, particularly in road vehicles. The conventional methods of speed control are to use a series resistance (or a parallel diverting resistance) which dissipates any unwanted power. But the battery life per charge can be extended by 40%, in a typical case, if power transistors or thyristors arc uscd as the controllable sources. This improvement with the use of semiconductor devices has

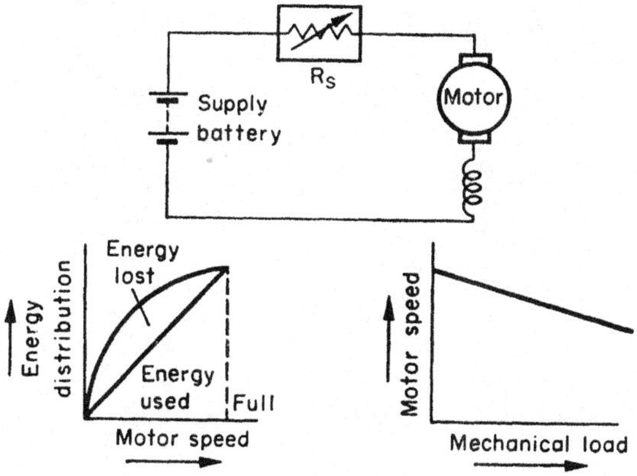

FIG. 11.22. D.C. motor control with series resistance.

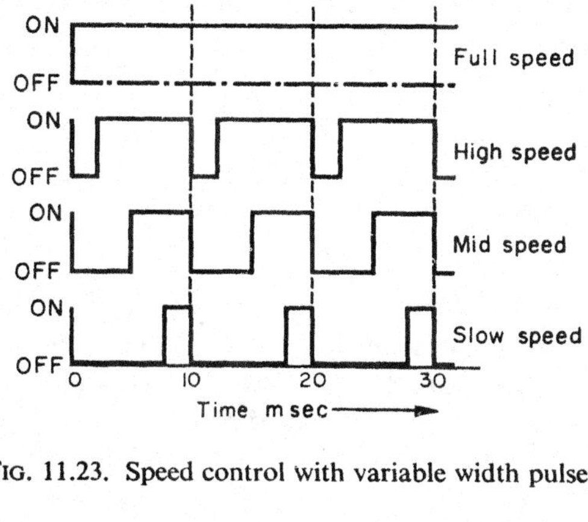

FIG. 11.23. Speed control with variable width pulses.

recently widened the applications of electric vehicles and reintroduced the possibility of electric cars as local runabouts.

The use of a series resistor to control the speed of a d.c. motor is particularly applicable witn a constant voltage source, such as a storage battery. This arrangement is shown in outline in Fig. 11.22, together with graphs indicating the energy and speed characteristics. This figure shows that the resistance control is inefficient at mid-speed and also suffers from poor regulation of speed with varying loads. The two graphs are drawn for series wound motors.

If a periodically operated switch is put in series with the motor, then the continual opening and closing of this switch may regulate the energy input. This is indicated by Fig. 11.23, where the

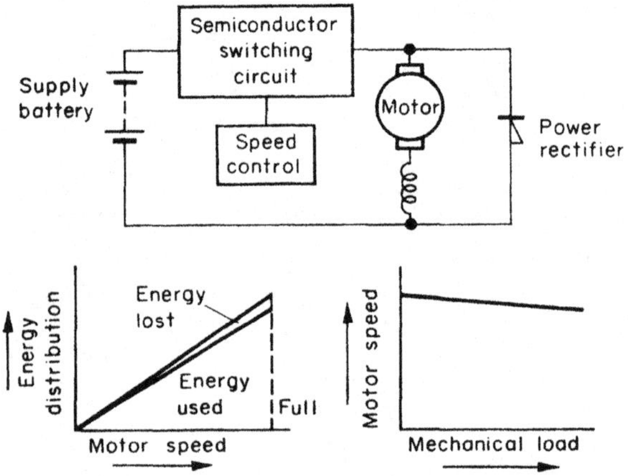

FIG. 11.24. D.C. motor control with semiconductor switch.

switching rate is at 100 Hz. Variation of the ON–OFF ratio provides the control of motor speed. Since the winding of the motor appears as a large inductance, so the pulsed input is smoothed, and the motor receives effective direct current. This switching control system with approximate energy and speed characteristics is sketched as Fig. 11.24.

The semiconductor switching circuit may incorporate either germanium power transistors or silicon thyristors. The associated control circuits are more simple when transistors are employed and the power dissipation problems are less with thyristors.

Figure 11.25 gives a more detailed block schematic of a possible arrangement, using transistors. In this arrangement the speed control element is a variable resistor in a form of resistance–capacitance oscillator. The first four units shown in this figure

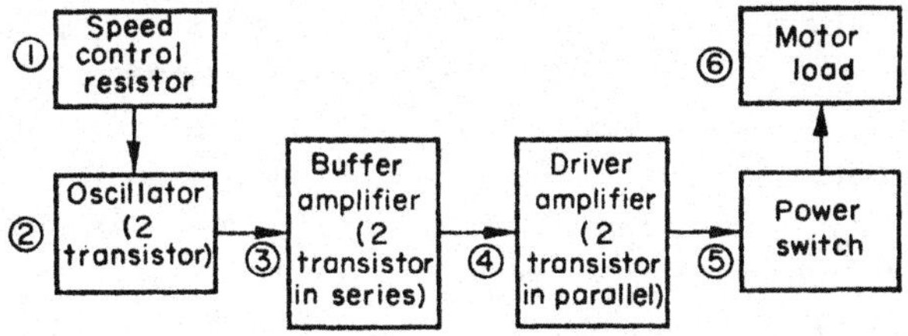

FIG. 11.25. Block schematic of motor control circuit with transistors.

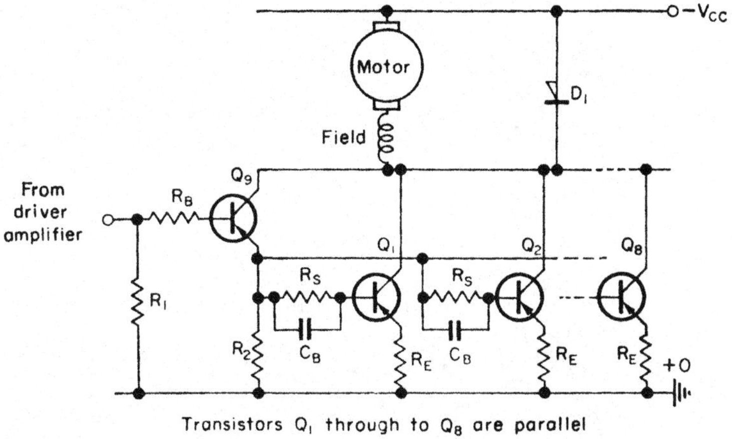

FIG. 11.26. Power switch, with parallel transistors, and motor load.

would be common for a wide range of motor sizes. The buffer and driver amplifiers are operating as emitter followers, with a common collector resistor, of the order of 1 Ω. In the power switch the number of germanium power transistors to be joined in parallel will depend on the motor rating. In one circuit for a ½ h.p., 24 V motor, four parallel driving transistors were needed to handle peak currents of up to 100 A (which occur on starting or stalling).

Details of a power switch, suitable for a 1 h.p. motor, and the motor load are in Fig. 11.26. Similar power switches may be joined in parallel and supplied from a common drive equipment, up to at least 5 h.p. rating. The base resistors R_B would be approximately 2 Ω and 2 W rating. The current-sharing emitter resistors R_E would be of the order of 0·1 Ω and 3 W rating. A large common heat sink is essential for the power transistors. An input capacitor (typically 8 μF) at the base of each transistor, improves the sharpness of the switching. A rectifier must be connected across the motor winding to minimize the possibility of damaging the transistors when the circuit is switched off. A large voltage pulse can be induced due to the inductance of the motor winding. The shunt rectifier reduces any such voltage pulse to the value of the supply voltage, which, in turn, must be within the minimum collector–emitter voltage rating of the transistors.

An overall power efficiency of nearly 95% has been claimed for this circuit arrangement during an average operating cycle.

Questions for Chapter 11

Q.11.1. With the aid of a block schematic, outline a method by which a circuit employing power transistors may control a d.c. traction motor. Point out the advantages of such a system.

Discuss the possible use of thyristors (S.C.R.) in a similar circuit and indicate the advantages, or otherwise, compared with transistors.

Q.11.2. Discuss the advantages of electronic switching, compared with electro-mechanical switching, in telephone exchanges.

Draw up a table to compare the relative properties of the available electronic devices.

Q.11.3. Compare critically the general features of amplitude and frequency modulation in communication systems.

Derive an expression for an amplitude-modulated carrier and explain how the original signal may be recovered in a diode detector. Graphically, or otherwise, show how distortion may be introduced by the detector stage.

(I.E.E., Nov. 1962)

Q.11.4. A point-contact diode gives an output current of $i = \alpha v + \beta v^2$ when biased to a certain portion of the forward characteristic, where i is in milliamps and v in volts. Show algebraically that this diode may be used for (a) amplitude modulation and (b) frequency changing of an amplitude-modulated signal.

Draw a circuit for such a diode suitable for the detection of a signal with a carrier frequency of 200 kHz. State appropriate values for the resistors and capacitors used.

Q.11.5. Explain in detail how a diode, in series with a parallel R.C. circuit, acts as a detector of A.M. voltages.

What is meant by the " detection characteristics " or dynamic characteristic of a diode detector? Explain how these may be measured in a laboratory and how they may be used to predict the performance of a diode detector operating in given conditions. (I.E.E., Dec. 1959)

PART III

Experiments with Semiconductors

CHAPTER 12

Laboratory Demonstrations: Semiconductor Characteristics and Circuits

THIS chapter describes circuits that may be used to show the characteristics of semiconductor devices and the operation of simple circuits. In general these circuits are not suitable for accurate measurements but would prove useful for lecture demonstrations.

12.1. Comparison of Metal and Semiconductor Diodes

Figure 12.1 shows a circuit which may be used to compare the d.c. characteristics of four common types of rectifier: copper oxide, selenium, germanium and silicon.

The equipment required includes a mains isolating transformer, variac, oscilloscope (preferably with a 19 in. screen) and a small circuit board. This board contains two resistors (of 500 Ω and 5 kΩ) and the four diodes under test. The diodes should be capable of passing about 25 mA in the forward direction and withstanding about 120 V in the reverse direction.

As indicated in the circuit of Fig. 12.1, the voltage across the rectifier is connected to the X input of the oscilloscope while a voltage proportional to the current flowing is fed to the Y input. The resulting trace on the C.R.O. screen shows the current/voltage characteristic of the rectifier. If the oscilloscope has a screen with reasonably long persistence the characteristics of the various rectifiers may be compared.

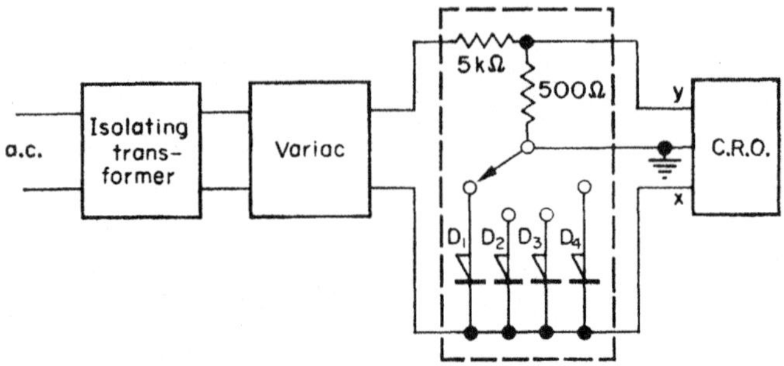

FIG. 12.1. Circuit to demonstrate rectifier characteristics.

12.2. Transistor Characteristics

12.2.1. *Grounded base*

Input, transfer and output characteristics of a transistor in the grounded base configuration may be demonstrated using the circuits shown in Figs. 12.2, 12.3 and 12.4. For convenience the same basic test unit is used. The ancillary equipment includes a large-screen oscilloscope, d.c. bias supplies, d.c. amplifier and a low frequency oscillator operating at about 1 Hz.

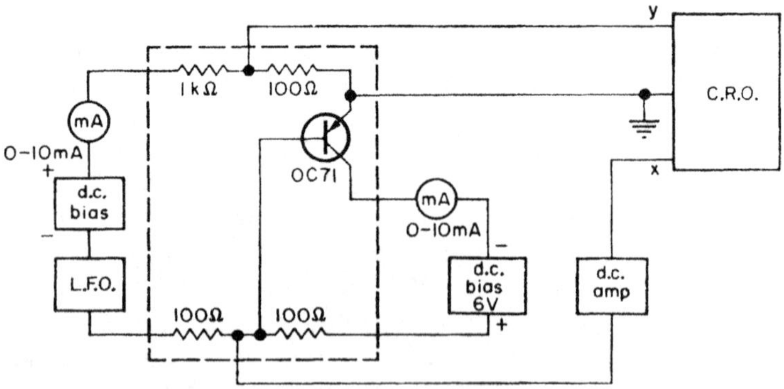

FIG. 12.2. Circuit to demonstrate transistor input characteristic (grounded base).

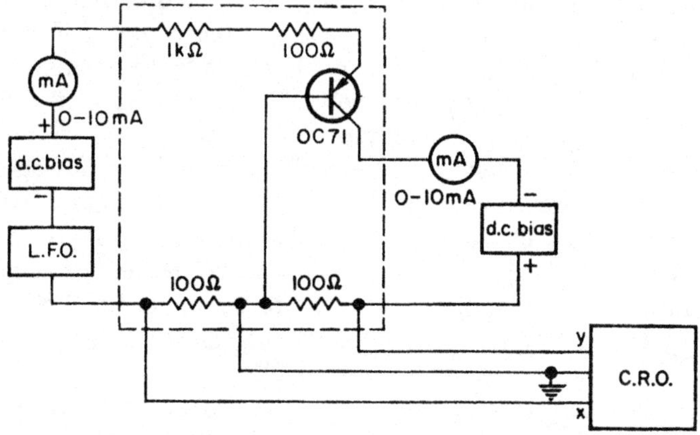

FIG. 12.3. Circuit to demonstrate transistor transfer characteristic (grounded base).

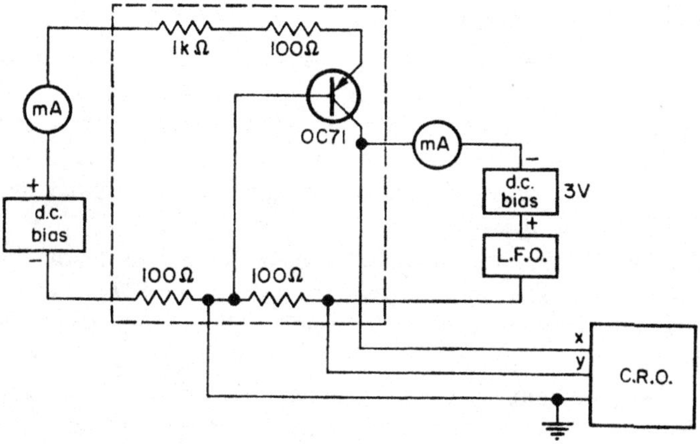

FIG. 12.4. Circuit to demonstrate transistor output characteristics (grounded base).

The input characteristic is displayed by earthing the emitter, connecting the base to the *X* input of the oscilloscope and applying the voltage across a series resistor to the *Y* input. The horizontal deflection is then a measure of voltage between emitter and base and the vertical deflection proportional to the emitter current. The effect of varying the collector voltage may be shown.

Figure 12.3 is a slight modification of Fig. 12.2. In this case the base of the transistor is earthed and voltages proportional to emitter and collector currents applied to the *X* and *Y* inputs of the oscilloscope. The phases of these currents are such that many oscilloscopes display the transfer characteristic in the first quadrant. If necessary a d.c. amplifier (giving unit gain and phase inversion) may be inserted before the *X* input.

To show the output characteristic, the oscillator is connected in the collector circuit, as shown in Fig. 12.4. The variation of collector current with collector voltage, keeping constant emitter current may then be displayed. By varying the emitter current a family of output characteristics may be obtained.

12.2.2. *Grounded emitter*

Due to the low values of base current, which have to be controlled, the circuits used for displaying the characteristics with grounded emitter are slightly more complex than in the previous case. The same display unit may be used for each characteristic and a similar range of ancillary equipment is required as for grounded base operation.

Figure 12.5 shows the circuit used to demonstrate the input characteristic. The voltage across the resistor of 3 kΩ is a measure of the current and provides the vertical deflection. The voltage between base and emitter is connected through a d.c. amplifier, which gives a gain of about ten times, to produce the horizontal deflection. Non-alignment of forward and return strokes may be counteracted by the phase shift introduced by putting a capacitor of 8 μF across the 3 kΩ resistor. This capacitor also serves to reduce 50 Hz ripple, which may be troublesome.

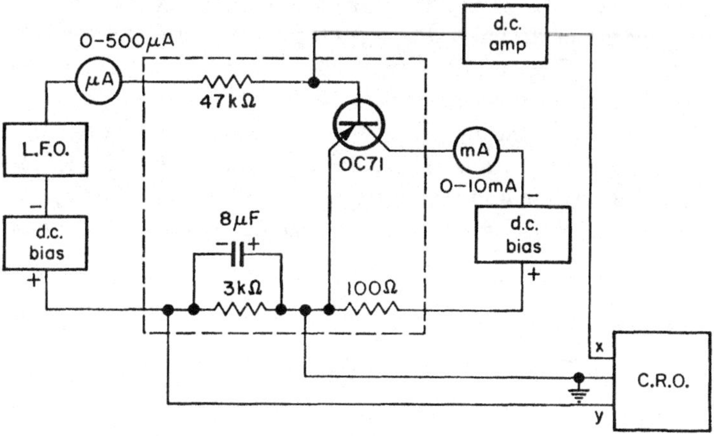

FIG. 12.5. Circuit to demonstrate transistor input characteristic (grounded emitter).

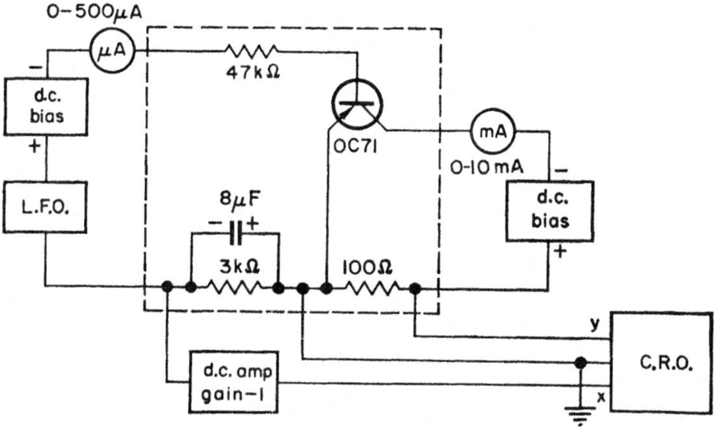

FIG. 12.6. Circuit to demonstrate transistor transfer characteristic (grounded emitter).

A slight modification of Fig. 12.5 is shown in Fig. 12.6. Here a voltage proportional to the base current is inverted in phase and connected to the *X* input of the oscilloscope. At the same time a voltage proportional to the collector current provides the vertical deflection. The resultant trace shows the transfer characteristic for the grounded emitter mode of operation.

In Fig. 12.7 the low frequency oscillator is connected in the collector supply lead. The collector voltage provides the horizontal deflection and the voltage across a resistor in series gives a vertical deflection proportional to the current. As in the case of

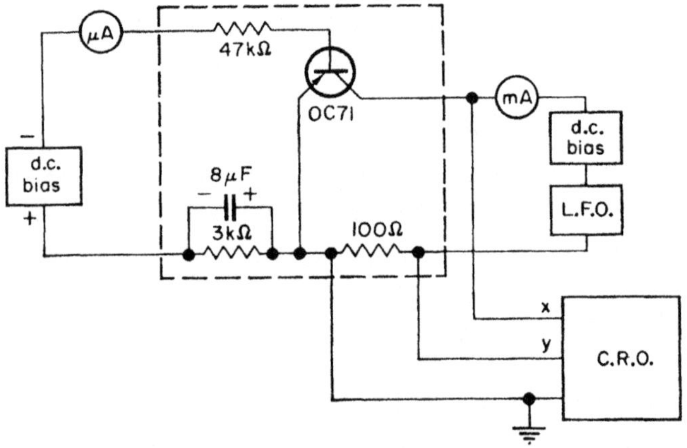

FIG. 12.7. Circuit to demonstrate transistor output characteristics (grounded emitter).

the common base configuration, if the oscilloscope has a long persistence screen, then a family of output characteristics may be clearly seen.

12.3. Characteristic of a Silicon Controlled Rectifier

The circuit used to display the current/voltage characteristic of an S.C.R. is shown in Fig. 12.8. This enables the static characteristic of the device and the effect of varying gate current to be displayed. A mains isolating transformer, oscilloscope, variac, bias voltage supply, milliammeter and special test unit are required.

The characteristic is displayed by feeding a fraction of the voltage across the S.C.R. to the *X* plates of the oscilloscope and a

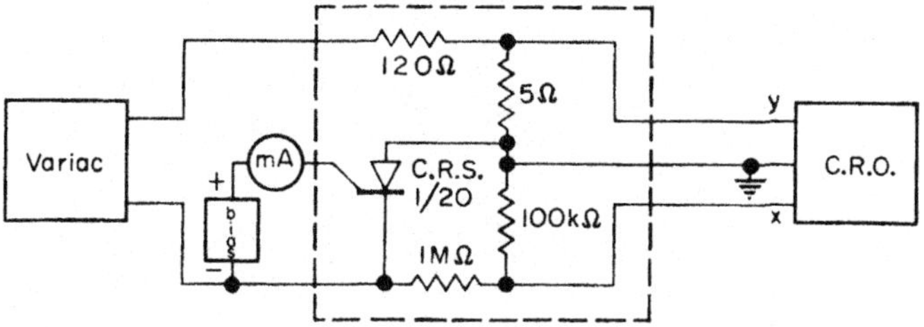

FIG. 12.8. Circuit to demonstrate thyristor characteristic.

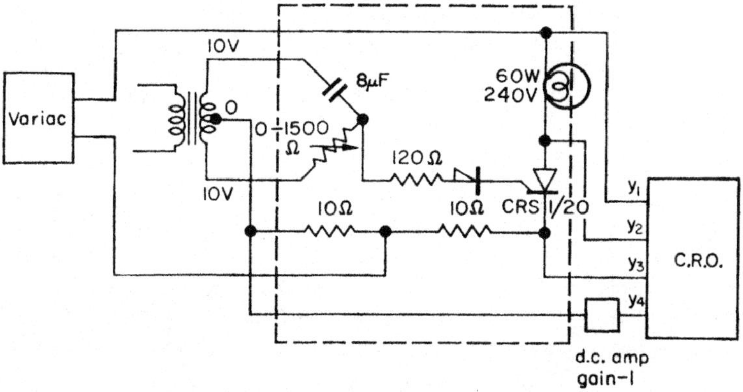

FIG. 12.9. Circuit to show phase control of a thyristor.

voltage, proportional to the current flowing, to the *Y* plates. The resulting trace shows the rapid increase in current as the rectifier starts to conduct and the reduced voltage required as the gate current is increased. For fine control of the base current a resistance in series with the bias supply may be required.

12.4. Phase Control of a Silicon Controlled Rectifier

In this demonstration an oscilloscope having four beams is used to show the supply voltage, the rectifier volt drop, the gate current and load current. The complete circuit is in Fig. 12.9, and includes a step-down transformer, a variac, a d.c. amplifier

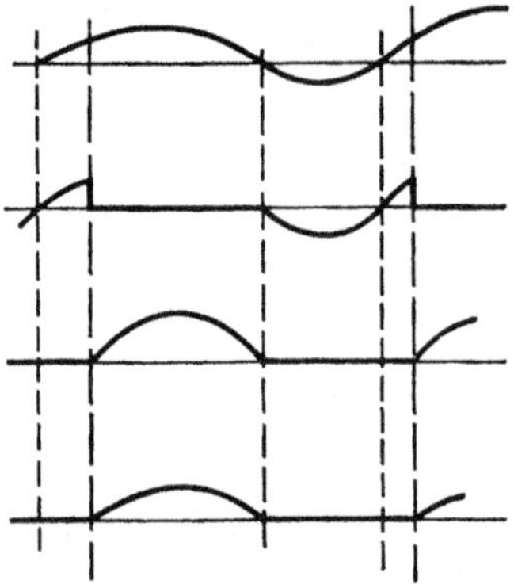

FIG. 12.10. Typical waveforms from a controlled rectifier.

and a special circuit board. This board contains four resistors, one capacitor, a diode, the controlled rectifier and a 60 W lamp, which provides a convenient load. Using an S.C.R. rated at 1 A and a supply of about 100 V from the variac, the current flow may be controlled from about zero to half a cycle (180°). Typical waveforms are shown in Fig. 12.10. The supply and rectifier voltages are connected directly to the oscilloscope. The current waveforms are represented by the voltages across two 10 Ω resistors.

12.5. Characteristic of a Zener Diode

A simple circuit which may be used to show the current/voltage characteristic of a small zener diode is given in Fig. 12.11. An

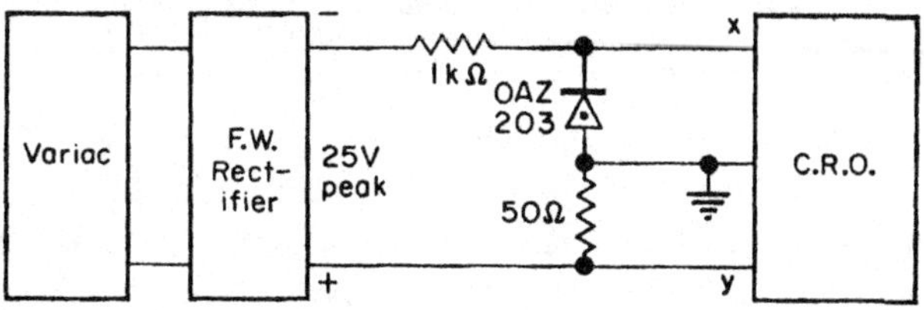

FIG. 12.11. Circuit to demonstrate characteristic of a zener diode.

unsmoothed full-wave rectified supply, of variable amplitude, two resistors and a demonstration oscilloscope are needed.

The horizontal axis of the trace, which is obtained on the C.R.O., is proportional to the voltage across the zener diode. The vertical deflection is a measure of the diode current. As the input voltage is increased there is negligible reverse current until the zener voltage is reached. The rapid increase in reverse current at the breakdown voltage of the diode may be clearly seen.

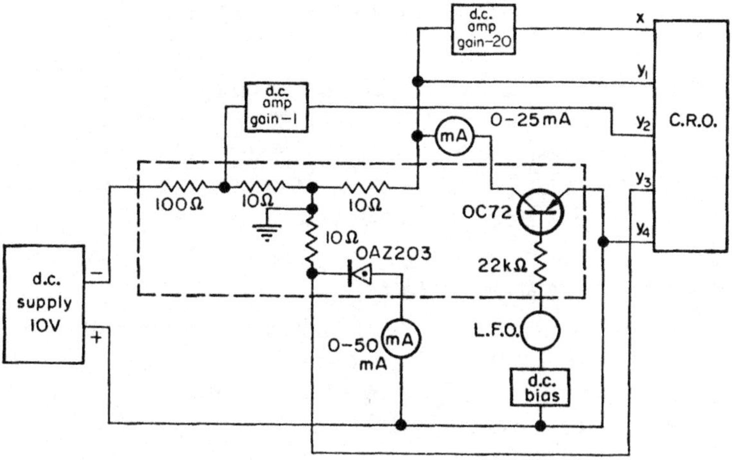

FIG. 12.12. Circuit to show stabilizing action of a zener diode.

12.6. Stabilizing Action of a Zener Diode

12.6.1. *Varying load current*

To show the voltage stabilizing action of a zener diode the circuit of Fig. 12.12 may be used. This circuit is necessarily complex, in order to show load, diode and supply currents and the voltage across the load. Two d.c. amplifiers, two d.c. supplies, a low frequency oscillator, two milliammeters and the four-beam oscilloscope are required. The special test unit includes five resistors, a small transistor and the zener diode. The output voltage will be stabilized at the zener voltage of the diode. By

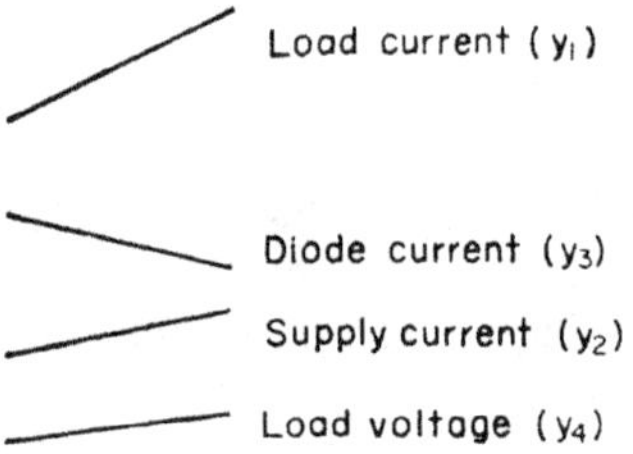

FIG. 12.13. Oscilloscope traces from a zener diode stabilizer.

suitable control of the base current, through the transistor, the current may be varied from about 0–20 mA. A voltage proportional to this swing in current provides the horizontal deflection for the oscilloscope. Typical oscilloscope traces for a supply voltage of 10 V are shown in Fig. 12.13.

12.6.2. *Varying supply voltage*

This demonstration involves a slight modification to the circuit shown in Fig. 12.12. The low frequency oscillator is connected in series with the supply to be stabilized and the internal time-base of the oscilloscope is used.

The supply voltage may then be varied by about $\pm 10\%$ from the nominal value. The traces obtained on the screen show that the variation in supply current is balanced by a variation in diode current, thus giving an almost constant load current. Alternatively, the positive side of the supply may be earthed, and the variation in input voltage compared with the variation in output voltage.

12.7. Demonstration of Hall Effect

The Hall effect may be demonstrated, and the mobility of charge carriers calculated, using the circuit of Fig. 12.14. This

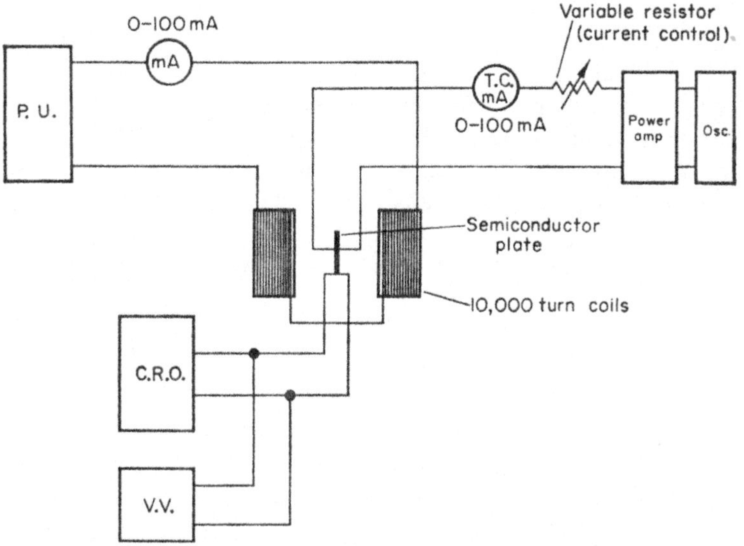

FIG. 12.14. Circuit to demonstrate the Hall effect.

demonstration uses a commercially available Hall Effect Probe and requires a small magnetic field. A power amplifier is required, in addition to an oscillator to give currents of up to 100 mA through the indium arsenide semiconductor plate. A d.c. power supply is used to give a magnetic field of the order of 10–20 mtesla. In addition a sensitive valve-voltmeter, to measure the Hall voltage, is essential, and a visual demonstration can be given with an oscilloscope in parallel with the voltmeter.

The effects of changing magnetic field and of varying magnitude and frequency of the current through the semiconductor may be clearly shown. If the dimensions of the semiconductor slice are known, then measurement of magnetic field strength, current flowing through the slice and Hall voltage, enable the Hall coefficient to be calculated from the equation

$$R_H = \frac{3\pi}{8}\left(\frac{E_H}{BJ}\right).$$

A voltmeter across the plate enables the resistance to be measured, and hence the resistivity to be estimated. The mobility of the charge carriers may then be calculated, since $\mu = R_H/\rho$ (as shown in Chapter 1).

An extension of the demonstration is to pass an alternating current through the field coils and a different frequency through the plate. The Hall voltage output is then an amplitude modulated wave, since the circuit behaves as a balanced modulator.

CHAPTER 13

Experiments with Semiconductor Circuits

THIS chapter describes a series of experiments suitable for students taking a course devoted to transistor technology at the Higher National Certificate level. Except for the first, these are all circuit experiments, since it is felt that characteristics are better shown by demonstration.

13.1. Variation of Current Gain with Frequency and Emitter Current

For this experiment a Phasemeter that also measures relative magnitudes was used. An oscillator, power supplies, milliammeters and an easily constructed test unit complete the circuit as shown in Fig. 13.1.

The a.c. equivalent for the circuit shown in Fig. 13.1 is given in Fig. 13.2 and from the circuit,

Common emitter current gain

$$\beta_0 = \frac{i_c}{i_b} = \frac{v_{OUT} \times 10^{-2}}{v_{IN} \times 10^{-6}} = \frac{v_{OUT}}{v_{IN}} \times 10^4 .$$

This is valid since the input impedance is very much less than 1 MΩ and 100 kΩ.

Also $$v_{REF} = v_{IN} \times 10^{-2}$$

giving $$\beta_0 = 100 \frac{v_{OUT}}{v_{REF}} .$$

Hence if the ratio v_{OUT}/v_{REF} is determined in magnitude and phase the current gain β_0 may be determined in magnitude and phase.

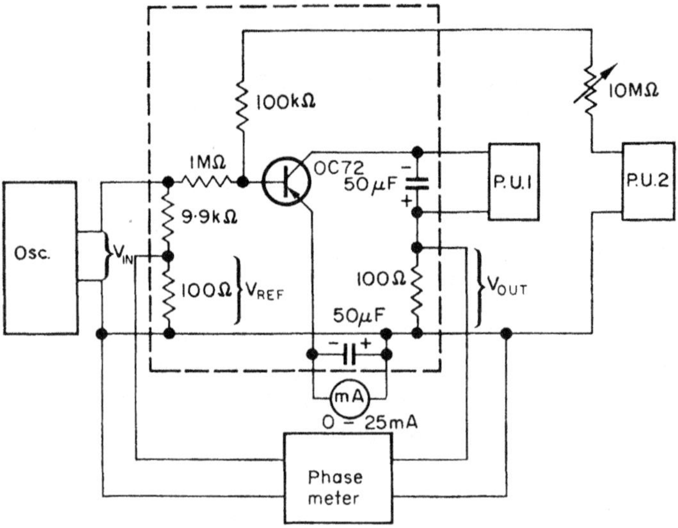

FIG. 13.1. Circuit to show the variation of β with frequency.

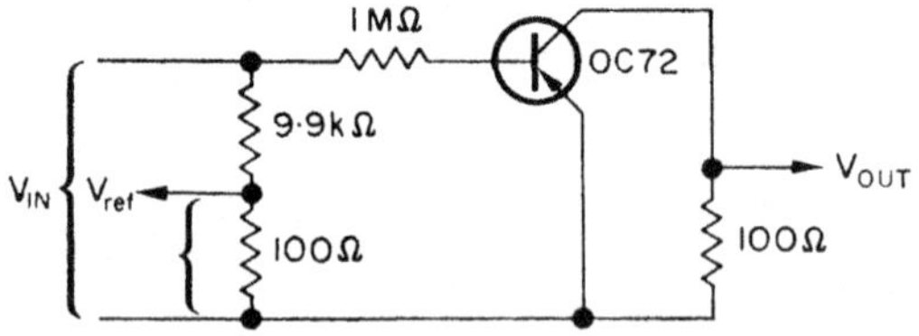

FIG. 13.2. Equivalent a.c. circuit of Fig. 13.1.

The theory given indicates the method of carrying out the experiment. The ratio of output to reference voltage may be measured over a range of frequency and emitter current and the current gain evaluated. Typical results are shown in Figs. 13.3,

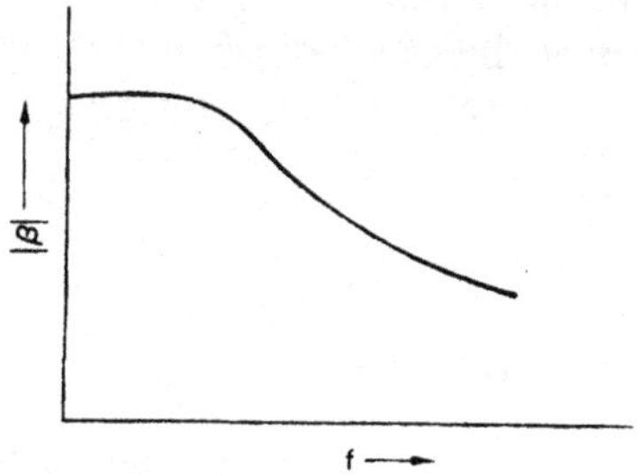

FIG. 13.3. Variation of magnitude of β with frequency.

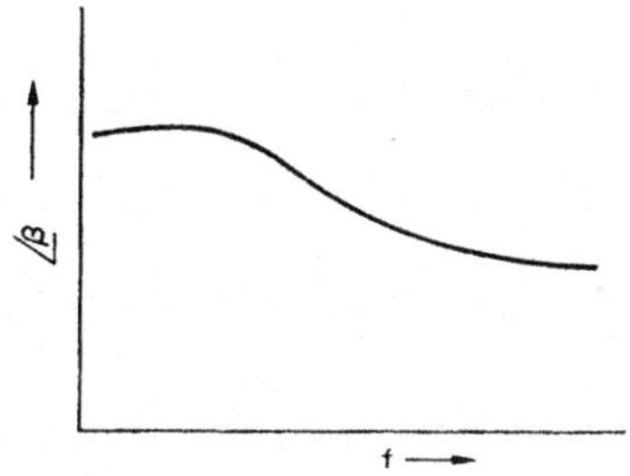

FIG. 13.4. Variation of phase of β with frequency.

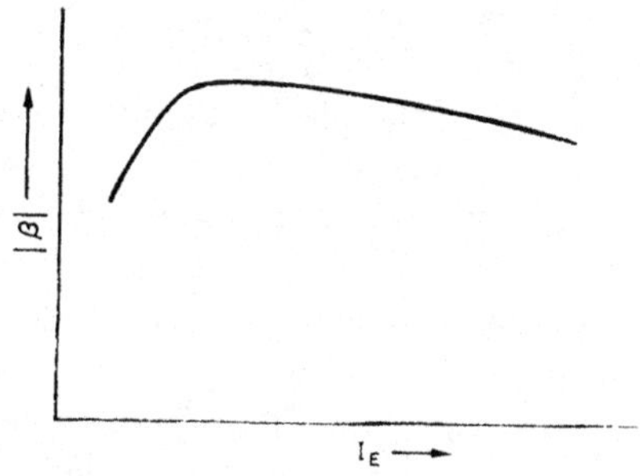

FIG. 13.5. Variation of magnitude of β with emitter current.

13.4 and 13.5. The emitter current, which may be as low as 10 μA for the initial part of the curve shown in Fig. 13.5, is controlled by the base current supply and the collector voltage is constant at -6 V.

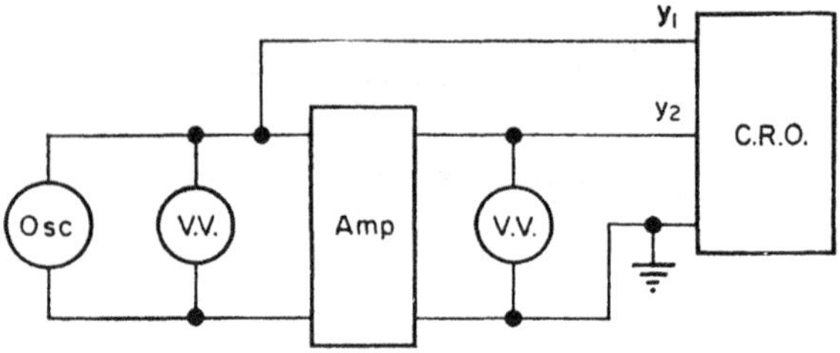

FIG. 13.6. Amplifier test circuit.

13.2. Common Emitter Amplifier

The test circuit for this experiment is shown in Fig. 13.6. The actual circuit used was the amplifier section of Fig. 13.10, but there are many suitable circuits available. Additional equipment includes two valve-voltmeters, an oscillator and a double beam C.R.O.

The experiment involves investigating the waveforms at various points in the circuit and plotting frequency response curves for single- and two-stage amplifier circuits. The cathode ray oscilloscope enables estimates of the phase response to be made and in particular the phase lead or lag at the 3 db points should be noted. It is interesting to note that the gain of a two-stage amplifier is not the square of that of a single-stage amplifier. This is due to the loading effect of the input impedance of the second stage on the first amplifying stage.

An extension of the experiment is to plot a curve showing the amplitude linearity. This may be measured for a range of input levels until distortion as indicated by the C.R.O. is noticed.

13.3. The Emitter Follower

A suitable emitter follower circuit is shown in Fig. 13.7. This experiment involves measurement of the frequency response and input and output impedances.

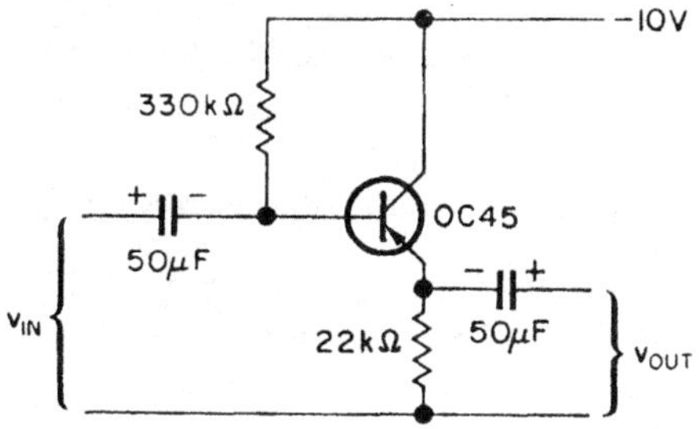

FIG. 13.7. Emitter follower circuit.

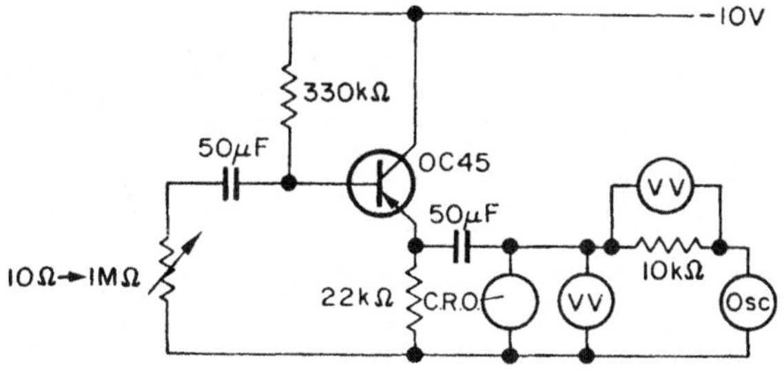

FIG. 13.8. Circuit to determine the output impedance of an emitter follower circuit.

Measurement of frequency response may be made with a circuit similar to that shown in Fig. 13.6 and typical results give a constant gain up to 100 kHz with the 3 db point at about 350 kHz.

The input impedance is conveniently measured by means of a *Q* meter and over the range 50 kHz to 2 MHz the input impedance is better than 100 kΩ in parallel with about 20 pF. These values correspond to a high impedance load but the effect of load variation may be easily determined.

Figure 13.8 shows a method of measuring the output impedance of the emitter follower. It ignores the angle of the output impedance, which is probably resistive and is a straightforward voltmeter–ammeter method of measuring impedance. A suitable level is about 10 mV at the emitter and typical results show an output impedance of between 200 Ω and 300 Ω over a frequency range 25 Hz to 250 kHz. This value assumes a low impedance source and the values are higher if the source resistance is increased.

13.4. The Tuned Collector Oscillator

A simple circuit that shows the principles of the tuned collector is shown in Fig. 13.9. The majority of the components may be

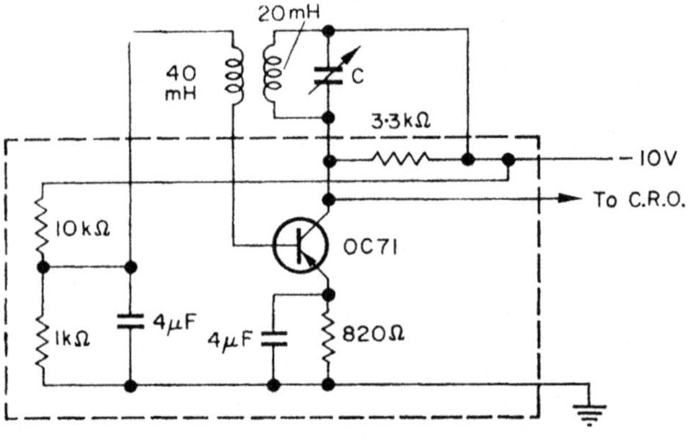

FIG. 13.9. Tuned collector oscillator.

mounted on a baseboard and additional equipment required includes a power supply, oscilloscope, a pair of inductances and a tuning capacitor. The 3·3 kΩ collector resistance simulates the load resistance.

A suitable experiment is to investigate the effect of change of tuning capacity, supply voltage and mutual coupling on the output frequency, amplitude and waveform. It is usually difficult to measure the coupling factor but graphs may be plotted to show the variation of frequency and amplitude with capacitance and supply voltage. Typical variations used in practice are 0·05 μF to 1 μF for the capacitance and 5–15 V for the supply voltage.

Further tests involve the production of a squegging output either by overcoupling or increasing the emitter decoupling capacitance. If a small crystal oven is available the effect of temperature may be observed. It is convenient to mount the transistor itself on the lid of a crystal oven.

Finally, under normal operating conditions the power efficiency may be estimated by measuring the d.c. input power and the power in the 3·3 kΩ load. The former is measured by voltmeter and ammeter while for the latter the output voltage is measured on the C.R.O.

13.5. Transistor R.C. Oscillators

This experiment uses two types of R.C. oscillator employing current feedback. The oscillator circuits are shown in Figs. 13.10

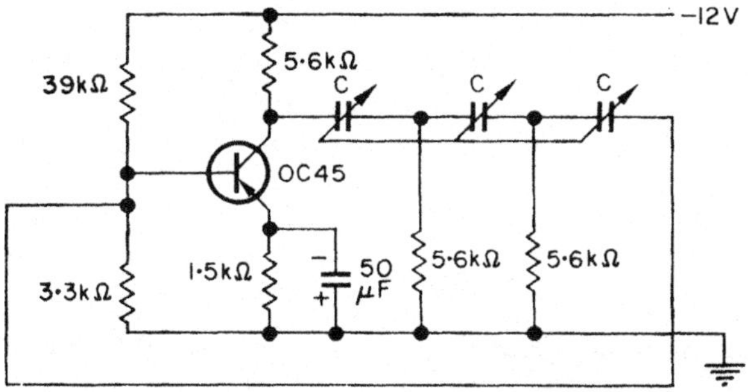

FIG. 13.10. Single-stage R.C. coupled oscillator.

and 13.11 and ancillary equipment includes a power unit, an oscilloscope and a frequency counter.

The experiment involves investigating the waveforms at various parts of the circuit and plotting curves relating frequency f and capacitance C. The latter may be varied from about 0·05 μF to 1 μF in the first circuit and from 0·001 μF to 1 μF in the second.

It is interesting to compare the voltage gain of the circuit at various frequencies of oscillation since this emphasizes the fact

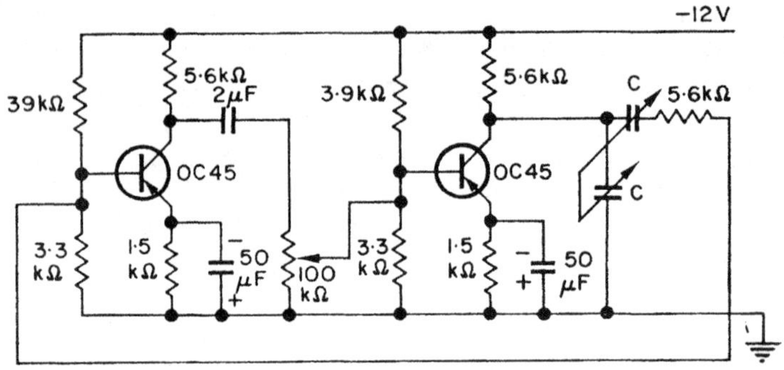

FIG. 13.11. Two-stage R.C. coupled oscillator.

that current gain is the important criterion in these oscillators. This is particularly noticeable in the first circuit where the frequency range is limited and transistors with a low value of current gain will not be suitable. Three stages of amplification may be required.

13.6. Multivibrator Circuits

This experiment involves a study of the operation of free running, monostable and bistable circuits. For the free running multivibrator shown in Fig. 13.12 waveforms at various points in the circuit may be observed and the effect of varying the base resistors noted. One or both of the base resistors may be reduced to about 7 kΩ and graphs plotted showing the variation in mark to space ratio and frequency of operation.

The circuit shown in Fig. 13.12 may be easily modified to give a monostable circuit by the addition of a small positive bias,

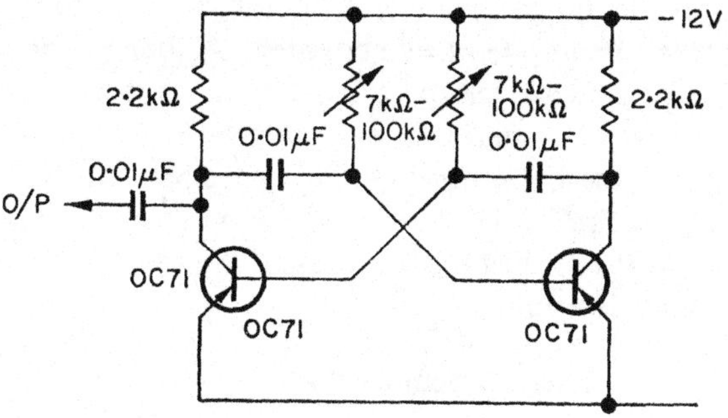

FIG. 13.12. Astable multivibrator.

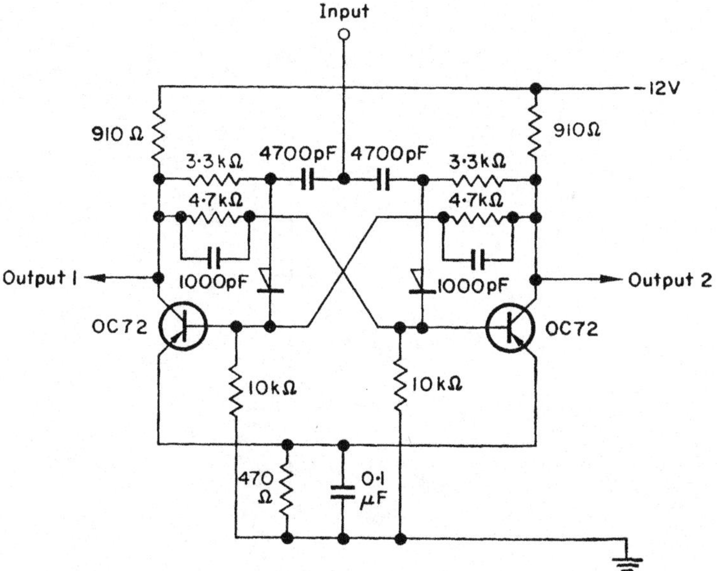

FIG. 13.13. Bistable multivibrator.

through a high resistance (10 kΩ), to one of the bases. If an oscillator is connected to the other base the circuit then acts as a pulse shaper and amplifier.

Figure 13.13 shows a bistable circuit that could conveniently be incorporated in a scale of two counter. A square wave input will be divided by two over quite a large range of frequency. The frequency range of the circuit may be determined and the effect of varying the base–collector coupling capacitors noted.

As an extension of the experiment the astable multivibrator may be coupled into the bistable circuit and the frequency range of the combination determined.

13.7. A Semiconductor Modulator

Figure 13.14 shows a simple modulator circuit which may be easily constructed. It is designed to operate at audio frequencies. Ancillary equipment such as audio frequency oscillators, valve voltmeters and a C.R.O. are usually available.

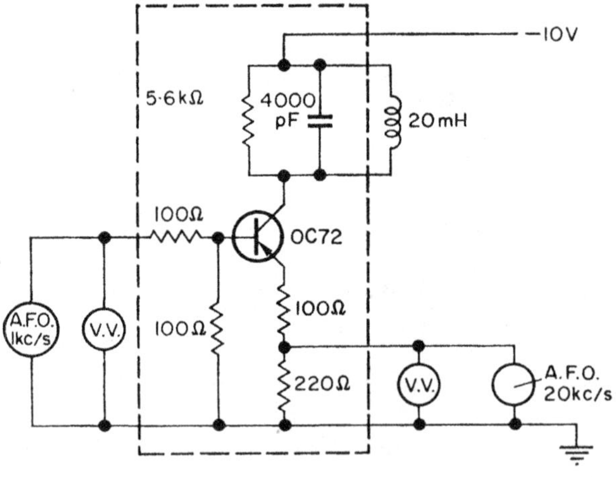

FIG. 13.14. Simple transistor modulator.

The optimum conditions for the above circuit give an unmodulated output of about 2·5 V r.m.s., and a 1 kHz input of about 0·3 V r.m.s. produces about 80% modulation. Curves may be plotted showing how the depth of modulation varies with the amplitude of the modulating frequency, i.e. amplitude linearity, and how it varies with modulating frequency, i.e. frequency linearity.

The experiment may be extended to cover a complete system comprising the modulator, an amplifier and detector as shown in Fig. 13.15. The amplifier output of many commercial valve voltmeters is suitable for the system and overall response curves may be determined.

FIG. 13.15. Simple transmission system.

CHAPTER 14

Design Experiments

THE six experiments suggested in this short chapter are intended as small-scale projects, to occupy not more than six hours each. Any student who attempts these designs will need to use some of the information given in preceding chapters of this book and also to apply a certain amount of thoughtful care. A suggestion for some suitable apparatus, but by no means an exhaustive list, is given at the end of each section.

14.1. Single-stage Audio Amplifier

(a) You are given three *pnp* transistors. By d.c. measurement of two input and two output characteristics, in the common emitter mode, decide which is the best transistor for an audio amplifier to use in a moderate temperature range. For these tests limit the base current to 100 μA and the collector emitter voltage to 10 V. State and discuss the reasons for your choice of transistor.

(b) We wish to build a single-stage resistance–capacitance coupled amplifier, with a good response over the whole audio frequency range, and with a stable low-distortion performance. Design this circuit for a 10 V battery supply and calculate suitable component values.

(c) Test your audio amplifier design. You may modify the circuit to get an optimum result, but make a clear note of what changes are made.

(d) Comment on the results, and discuss where such a circuit would be useful. Point out the limitations of your design.

(e) Two small germanium transistors, chosen for widely differing properties, and one silicon transistor, with relatively low gain, are suitable for this experiment. These can be mounted on a small Perspex base. A double-beam general purpose oscilloscope, an electronic voltmeter, with d.c. scale, and two multimeters, should be available for use. A variable d.c. supply is also essential.

14.2. Simple D.C. Stabilizer

(a) Measure the current/voltage characteristic of a small silicon junction zener diode. Limit both forward and reverse current to 50 mA.

(b) Investigate the variation of zener voltage (which may be defined as at a reverse current of 5 mA) with temperature up to 100°C.

(c) Design a simple d.c. voltage stabilizer circuit with this zener diode and a series resistance. You may assume input voltage variations of $\pm 10\%$, with any suitable supply voltage, and a load current which may be in the range of 20–50 mA.

(d) Test the operation of your stabilizer circuit. From the resulting graphs calculate the stabilization, output resistance and temperature stability. Compare these results with the same parameters derived from approximate calculations.

(e) Discuss the limitations and possible applications of this simple d.c. stabilizer circuit.

(f) A zener diode in the Mullard OAZ 200 to 213 range is very suitable. Two resistance boxes, capable of carrying currents up to 100 mA, are required and a well-stabilized d.c. power supply is essential.

14.3. Power Stabilizer Circuits

(a) You are given the three following semiconductor devices on a small aluminium chassis, with the approximate characteristics as quoted:

Silicon zener diode, with a zener voltage of 6·4 V (at a reverse current of 5 mA) and a reverse resistance of 2 Ω in the current range from 5 mA to a maximum of 50 mA.

Germanium *pnp* power transistor, with a current gain of $\beta = 40$ at a collector current of 1 A. The maximum power dissipation is 6 W at 50°C, as mounted, and the maximum collector current is 5 A.

Small *pnp* transistor, with a current gain of $\beta = 50$, at a collector current of 20 mA. The maximum power dissipation is 100 mW at 50°C and the maximum collector current is 100 mA.

(b) Design two d.c. stabilizer circuits to supply a load which may vary from 80 Ω to 15 Ω. One circuit should have the power transistor in shunt and the other have the power transistor in series. The input voltage may change within the range from 15·5 V to 8·5 V. You may ignore temperature variations.

(c) Measure the performance of your stabilizer circuits, and modify circuit resistance values to improve the performance, if possible.

(d) Explain how the better circuit may be further improved to allow for changing conditions. Compare the performance with that of the simple stabilizer in Section 14.2.

(e) Two resistance boxes capable of carrying currents to 1 A are needed, one box incorporating 0·1 Ω steps of resistance. A third resistance box rated to 100 mA is also required. A well-regulated source of direct current, giving at least 1 A, is essential.

14.4. Frequency Modulation, with a Varactor

(a) Measure the junction capacitance of a varactor diode, with reverse voltages from 0 V to −20 V, at room temperature and 50°C.

(b) Attempt to obtain frequency modulation of an inductance–capacitance oscillator, at a carrier frequency of 470 kHz. Use both d.c. and audio frequency signals.

(c) Investigate the performance of your F.M. circuit. Compare and contrast this frequency modulator with other methods of obtaining F.M.

(d) Standard transistor circuit boards are useful to provide a 470 kHz oscillator and I.F. amplifier. A frequency counter and tuned detector are also needed. The DD51 series of junction diodes, manufactured by G. & E. Bradley (a division of Joseph Lucas Ltd.) are very suitable for this experiment.

14.5. Semiconductor Gate Circuits

(a) You are given three each of two types of diodes. Check the forward and reverse d.c. characteristics quickly. Limit the reverse voltage to 50 V and the forward current to 10 mA.

(b) Design a three-input AND gate, using the better diodes. Use a 6 V supply and assume the input pulses will be +2 V. Check the validity of your design, and sketch the output waveforms under various input conditions.

(c) Using the other three diodes, within the same voltage conditions, construct a three-input OR gate. Explain, from your experimental results, how the requirements for the OR gate differ from those of the AND gate.

(d) Construct a three-input NOR gate circuit employing a single transistor, again within the voltage conditions of 6 V supply and +2 V signal. Note input and output waveforms.

(e) Comment on your results, paying attention to the various output waveforms. Discuss the properties of the various circuits and indicate the possible fields of application.

(f) A small board with six diodes (three silicon junction and three germanium point-contact) and one transistor can be easily made. A 6 V dry battery and 2 V accumulator can provide the d.c. sources. Two waveform generators and a single-beam oscilloscope, with good frequency response and fine focus, are required.

14.6. Push-pull Driver Amplifier Stage

(a) Two transistors which are, nominally, a complementary pair, are supplied for this experiment. Measure a few collector current/collector–emitter voltage characteristics for the *pnp* and *npn* transistors.

(b) Design an audio frequency push-pull amplifier stage using these two transistors, driven from a stage with a single transistor.

(c) Measure the gain/frequency response and the distortion content given by the resulting push-pull amplifier stage. Attempt to reduce the distortion by the introduction of negative feedback. Estimate the overall power efficiency.

(d) Discuss the value and drawbacks of your circuit design and suggest possible lines of improvement.

(e) Low cost *pnp* and *npn* germanium transistors are available in the 2N1302 to 2N1309 series from Texas Instruments Ltd. An existing phase-splitting amplifier could be made available, so that attention may be concentrated on the complementary pair stage. Some accurate means of measuring second and third harmonic distortion is required.

Appendix A

Answers to Numerical Questions

A.1.1. 0·265 Å, −54·8 eV, 1·058 Å, −13·7 eV.

A.1.2. 0·00568 m^2/V sec, $5{\cdot}68\times10^{-11}$ m^3/C, 1·76 A/cm^2.

A.1.3. 5·2 Ω-m, 25 Ω-m, 0·534 Ω-m, 0·64 Ω-m.

A.1.4. $1{\cdot}56\times10^{20}$, 0·047 m^3/C, 0·135 Ω-m, 0·35 m^2/V sec.

A.3.4. (a) $\theta_{h(\max)} = 7{\cdot}5°C/W$.
(b) $\theta_{h(\max)} = 29°C/W$.

A.4.4. $h_{11} = 70\ \Omega$, $h_{12} = 10^{-3}$, $h_{22} = 10^{-6}$ mho, $h_{21} = -0{\cdot}98$.
$h'_{11} = 3500\ \Omega$, $h'_{12} = 2{\cdot}5\times10^{-3}$, $h'_{22} = 0{\cdot}5\times10^{-4}$ mho, $h'_{21} = 49$.
$h''_{11} = 1000\ \Omega$, $h''_{12} = 1$, $h''_{22} = 0{\cdot}5\times10^{-4}$ mho, $h''_{21} = -50$.

A.4.5. $Z_{11} = 1050\ \Omega$, $Z_{12} = 1000\ \Omega$, $Z_{22} = 1\ M\Omega$, $Z_{21} = 980\ k\Omega$.
$Z'_{11} = 1050\ \Omega$, $Z'_{12} = 50\ \Omega$, $Z'_{22} = 20\ k\Omega$, $Z'_{21} = -980\ k\Omega$.
$Z''_{11} = 1\ M\Omega$, $Z''_{12} = 20\ k\Omega$, $Z''_{22} = 20\ k\Omega$, $Z''_{21} = 1\ M\Omega$.

A.4.6. $Y_{11} = 0{\cdot}0143$ siemen, $Y_{12} = -1{\cdot}43\times10^{-5}$ siemen,
$Y_{22} = 1{\cdot}5\times10^{-5}$ siemen, $Y_{21} = -0{\cdot}014$ siemen,
$Y'_{11} = 2{\cdot}86\times10^{-4}$ siemen, $Y'_{12} = 7{\cdot}14\times10^{-7}$ siemen,
$Y'_{22} = 1{\cdot}5\times10^{-5}$ siemen, $Y'_{21} = 0{\cdot}014$ siemen,
$Y''_{11} = 2{\cdot}86\times10^{-4}$ siemen, $Y''_{12} = -2{\cdot}86\times10^{-4}$ siemen,
$Y''_{22} = 0{\cdot}05$ siemen, $Y''_{21} = -0{\cdot}05$ siemen.

A.5.2. Diode starts to conduct 0·78 msec after start of half-cycle, and stops conducting 5·97 msec after start of half-cycle.

A.5.5. Series resistor $R_B = 30\ \Omega$.
Power dissipated in R_B (with $R_L = 150\ \Omega$) = 0·5 W.
Diode current (when $R_L = 200\ \Omega$) = 42 mA.

A.5.7. $S = 0{\cdot}048$ or 4·8%.
$R_0 = 5{\cdot}7\ \Omega$.

A.5.8. Suitable component values are, approximately,

$R_B =$ 3 Ω and 3 W,

$R_S =$ 120 Ω and 1 W.

Zener diode with $V_Z = 6{\cdot}5$ V and maximum current = 100 mA.
Power transistor with $\beta = 50$ and maximum dissipation ⇒ 5 W.

A.6.2. $V_C = -5{\cdot}15$ V, $I_C = 1{\cdot}47$ mA.

A.6.4. 2000 Ω, 3·6 V (peak to peak), 1·9 mA (peak to peak).
500 Ω, 1·08 V (peak to peak), 2·15 mA (peak to peak).

A.6.5. 4·35 V (peak to peak), 42 times.

A.6.6. 200 kΩ.

A.7.1. (a) $R_D = 200\ \Omega$.
(b) $P_{OUT} = 3$ W.
(c) $I_{d.c.} = 100$ mA.
(d) V_{OUT} (peak to peak) $= 70$ V.

A.7.2. A.C. output power = 2 W.
Mean current taken = 144 mA.
Peak to peak output voltage = 62·5 V.

A.7.3. Suitable base bias resistor = 50 kΩ.
Approximate undistorted available output power = 60 mW.

A.7.4. Harmonic content $= \frac{5}{90}$ (or 5·5%).

A.7.5. Second harmonic percentage = 0·0011%.

A.8.1. Condition for oscillation is $Rr = \frac{\alpha_0 L}{C} - \frac{LC}{c_1{}^2}$.

A.8.2. Frequency of oscillation is $f = \frac{1}{2\pi\sqrt{(L_1 C)}}\sqrt{(1 + r/R)}$.

A.8.3. $f = \frac{\sqrt{6}}{2\pi CR}$, 17,800 Hz.

A.8.4. 14·1 Ω, −13·3 Ω.

A.8.5. 504 Mc/s, 186 MHz.

Appendix B

References for Further Reading

Chapter 1

SHOCKLEY, W., Transistor Physics, *Proc. I.E.E.*, Vol. 103, Part B, pp. 23–41, 1956.

Chapter 2

PRZYBYLSKI, J. and ROBERTS, G. N., The Design and Construction of Tunnel Diodes, *Journal Brit. I.R.E.*, December 1961.

Chapter 3

EVANS, J., *Fundamental Properties of Transistors* (2nd edition), Heywood, London, 1962.

KLEIMACK, J. J. and THEUERER, H. C., Epitaxy and Transistor Fabrication, *Bell Lab. Record*, September 1962.

WARNER, R. M., Epitaxial Growth and Devices, *Electronics*, May 18th, 1962.

HIBBERD, R. G., Transistors and Associated Semiconductor Devices, *Proc. I.E.E.*, Vol. 106, Part B, May 1959.

Chapter 4

ABRAHAMS, J. R. and COVERLEY, G. P., *Signal Flow Analysis*, Pergamon Press, Oxford, 1965.

Chapter 5

ASHBY, D., Metal and Semiconductor Rectifiers, *I.E.E. Journal*, Vol. 7, November 1961.

READ, J. C., Rectifiers and Rectifier Applications, *Proc. I.E.E.*, Vol. 110, No. 4, April 1963.

SCHADE, O. H., Analysis of Rectifier Operation, *Proc. I.R.E.*, Vol. 31, 1947.

PARKER, P, *Electronics*, ch. 14, Arnold, London, 1950.

CHANDLER, J. A., Characteristics of Zener Diodes, *Electronic Engineering*, February 1960.

SPENCER, R. H. and GRAY, T. S., Transistor Voltage Regulator, *Communication and Electronics* (*A.I.E.E.*), Vol. 75, p. 15, March 1956.

BELL, J. S. and WRIGHT, P. G., Stabilized Voltage Supplies using Transistors, *Electronic Engineering*, December 1960.

ORR, E. R., SMITH, R. and BUDEX, J. A., *Thyristors in A.C. Power Circuits*, Application Report MF/123, Standard Telephones and Cables Ltd. (Rectifier Division), Harlow, Essex, 1963.

Chapter 6

JOHNSON, L. B. and VERMES, Miss P., *D.C. Stabilization of Junction Transistors*, Mullard Tech. Comm. No. 23, April 1957.

JANSSON, L. E., *High Frequency Amplification using Junction Transistors*, Mullard Tech. Comm. No. 26, October 1957.

KEMHADJIAN, H., *Transistor Amplifiers for D.C. Signals*, Mullard Tech. Comm. No. 36, December 1958.

Chapter 7

HILBOURNE, R. A. and JONES, D. D., Transistor Power Amplifiers, *Proc. I.E.E.*, Vol. 102, Part B, p. 763, 1955.

MILNES, A. G., Transistor Power Amplifiers with Switched Mode of Operation, *Trans. A.I.E.E.*, Part 1, p. 368, July 1956.

TOBEY, R. and DINSDALE, J., Transistor Audio Power Amplifier, *Wireless World*, p. 565, November 1961.

PARKER, P., *Electronics*, Arnold, London, 1950 (see pp. 341–6 for discussion of Class C amplifiers).

Chapter 8

LIGHT, L. H. and HOOKER, P. M., Transistor D.C. Converters, *Proc. I.E.E.*, Vol. 102, Part B, pp. 775–86, 1955.

OAKES, F., Design Considerations for Junction Transistor Oscillators for the Conversion of Power from Direct to Alternating Current, *Proc. I.E.E.*, Vol. 104, Part B, pp. 307–17, 1957.

THOMPSON, P. M. and BATESON, J., An Introduction to the Tunnel Diode as a Circuit Element, *Journal Brit. I.R.E.*, July 1961.

CRIPPS, L. G., *Low Frequency Transistor Oscillators*, Mullard Tech. Comm. No. 22, March 1957.

Chapter 9

BEAUFOY, R. and SPARKES, J. J., The Junction Transistor as a Charge Controlled Device, *A.T.E. Journal*, 13, No. 4, p. 310, 1957.

EBERS, J. J. and MOLL, J. L., Large Signal Behaviour of Junction Transistors, *Proc. I.R.E.*, Vol. 42, p. 1767, 1954.

CHILTON, R. H., *Binary Counting Circuits using Alloy Junction Transistors*, Application Report MK/144, Standard Telephones and Cables Ltd., Footscray, Kent, 1962.

HARRIS, J. R., Direct-coupled Transistor Logic Circuitry, *Trans. I.R.E.*, Vol. EC7, No. 1, pp. 6–17, March 1958.

MILES, R. J., *The Design of a Standard Block for a Digital Computing System*, Mullard Tech. Comm. Vol. 4, No. 38, April 1959.

LINGWOOD, G. J. and GRIGG, R. H., An Instructional Aid for Digital Computer Logic, *The Radio and Electronic Engineer*, Vol. 25, No. 4, April 1963.

Chapter 10

SHAW, D. and CRUMP, B., Thermal and Optical Behaviour of Photo-Transistors, *Electronic Engineering*, December 1960.

PEARSON, G. L., Electricity from the Sun, *Bell Telephone Labs. Monograph* 2658 (also published in *Proceedings of the World Symposium on Applied Solar Energy*, pp. 281–8, April 1956).

SHIVE, J. N., The Properties of Germanium Photo-Transistors, *Journal of the Optical Society of America*, Vol. 43, p. 239, 1953.

SKETCH, H. J. H., Generation of Power in Satellites, *I.E.E. Journal*, Vol. 8, No. 87, March 1962.

Chapter 11

TERMAN, F. E., *Electronic and Radio Engineering*, McGraw-Hill, New York, 1955 (see pp. 523–617 for A.M. and F.M.).

REGISTRAR, J., Silicon Junction Devices as Variable Capacitors, *Electronic Engineering*, December 1961.

FORSTER, J. H. and RYDER, R. M., Diodes can do almost Anything, *Bell Lab. Record*, January 1961.

COOKE, P., Ferrite-core Stores, *I.E.E. Journal*, Vol. 9, August 1963.

FLOOD, J. E. and DELLER, W. B., The *pnpn* Diode as a Crosspoint, *Proc. I.E.E.*, Vol. 107, Part B, Supplement 20 (Conference on Electronic Telephone Exchanges) 1960.

Chapter 12

PRIDHAM, G. J., *Demonstrations and Experiments with Electronic Instruments*, Vols. 2 and 3, Airmec Ltd., High Wycombe, 1964.

Chapter 13

PRIDHAM, G. J., *Demonstrations and Experiments with Electronic Instruments*, Vol. 1, Airmec Ltd., High Wycombe, 1963.

Chapter 14

WILLIAMS, E. M. and YOUNG, F. J., *Electrical Engineering Problems*, McGraw-Hill, New York, 1960.

Appendix C

Proof of $\mu = \frac{r_e}{2r_c}(1-\alpha_0)$

From Fig. 4.1

$$v_{cb'} = i_c r_c + \alpha_0 i_e r_c$$

$$\simeq -i_e(1-\alpha_0)r_c, \quad \text{since } i_c \simeq -i_e.$$

For a small change $\delta v_{cb'}$ in $v_{cb'}$, let the corresponding change in i_e be δi_e.

Then $$\delta v_{cb'} = -\delta i_e(1-\alpha_0)r_c.$$

Also $$v_{eb'} = i_e r_e + \mu v_{cb'}$$

$$= i_e r_e - \mu i_e(1-\alpha_0)r_c.$$

If a change of δi_e in i_e produces a change $\delta v_{eb'}$ in $v_{eb'}$,

$$\delta v_{eb'} = \delta i_e r_e - \mu \delta i_e(1-\alpha_0)r_c.$$

A decrease in $v_{cb'}$ reduces the base width and leads to an increase in i_e and hence $v_{eb'}$

Therefore $$\mu = -\frac{\delta v_{eb'}}{\delta v_{cb'}} = -\frac{\{\delta i_e r_e - \mu \delta i_e(1-\alpha_0)r_c\}}{-\delta i_e(1-\alpha_0)r_c},$$

i.e. $$-\mu(1-\alpha_0)r_c = -r_e + \mu(1-\alpha_0)r_c$$

or $$\mu = \frac{r_e}{2r_c(1-\alpha_0)}.$$

Appendix D
Classification of Symbols

Symbols	*Meaning*
α	Current gain in common base.
$\bar{\alpha}$	Large signal direct current gain (common base).
α_0	Small signal, low frequency current gain (common base.
A	Cross-sectional area of a conductor (metre^2).
β	Current gain in common emitter.
$\bar{\beta}$	Large signal direct current gain (common emitter).
β_0	Small signal, low frequency current gain (common emitter).
$\boldsymbol{B}$	Flux density (tesla).
$\underline{\gamma}$	Current gain in common collector.
γ	Velocity of light *in vacuo* (3×10^8 metre/sec).
c	Large signal direct current gain (common collector).
C	Capacity (farad).
c_c, c_e	Collector, or emitter, capacity.
d	First sub-shell.
ϵ	Base of natural logarithms.
ϵ_0	Permittivity of free space $\left(\frac{1}{36\pi} \times 10^{-9} \text{ farad/metre}\right)$.
e	Charge on an electron ($1{\cdot}6 \times 10^{-19}$ coulomb).
E	Electric field strength (volt/metre).
E_1, E_2	Energy levels 1 and 2, respectively.
E_F	Fermi energy level.
E_H	Hall field strength (volt/metre).
f	Fourth sub-shell.
f_α, f_β	-3 db frequencies for α and β, respectively.
f_1	Frequency at which $\beta = 1$.
$f_{c.o.}$	-3 db frequency for current gain.
f_A	Parallel resonant (or anti-resonant) frequency of a crystal.
f_R	Series resonant frequency of a crystal.
g_m	Mutual conductance (amp/volt).
G_P	Power gain.
η	Power efficiency (usually percentage).
h	Planck's constant ($6{\cdot}624 \times 10^{-34}$ joule-sec).
h_{11}, h'_{11}, h''_{11}	A.C. input impedance, with output short-circuited (common base, emitter or collector).
h_{12}, h'_{12}, h''_{12}	A.C. voltage feedback ratio, with input open-circuited (common base, emitter or collector).

$h_{21}, h'_{21}, h''_{21}$	A.C. forward current gain, with output short-circuited (common base, emitter or collector).
$h_{22}, h'_{22}, h''_{22}$	A.C. output admittance, with input open-circuited (common base, emitter or collector).
θ	Thermal resistance (°C/watt).
θ_h	Thermal resistance, heat sink to ambient.
θ_i	Thermal resistance of insulation.
θ_m	Thermal resistance, mounting base.
i_b	Alternating base current.
i_c	Alternating collector current.
i_e	Alternating emitter current.
I	Moment of inertia of electron.
I	Direct current (amp).
I_B	Direct base current.
I_{BO}	Breakover current (in thyristor).
I_c	Direct collector current.
I_{co}	Transistor leakage current, with common base (emitter open-circuit).
I'_{co}	Transistor leakage current, with common emitter (base open-circuit).
I_E	Direct emitter current.
I_G	Gate current (in thyristor).
I_s	Reverse saturation current of a diode.
I_s	Sustaining current (in thyristor).
J	Current density (amp/metre^2).
k	Boltzmann's constant ($1{\cdot}38 \times 10^{-23}$ joule/degree abs.).
K	First electron energy shell.
K	Stability factor (transistor amplifiers).
K_T	Temperature stability factor.
λ	Wavelength.
l	Subsidiary quantum number.
L	Length of a conductor (metre).
L	Second electron energy shell.
L_s	Series inductance of tunnel diode.
μ	micron (10^{-6} metre or 10^4 Ångström).
μ	Voltage feedback factor.
μ_e, μ_h	Mobility of electron or hole (metre^2/volt sec).
m	Modulation index of A.M. (usually percentage).
m	Magnetic quantum number.
m	Electron rest mass ($9{\cdot}108 \times 10^{-31}$ kilogram).
m_f	Modulation factor in F.M.
m_s	Spin quantum number.
M	Third electron energy shell.
n	Principal quantum number.
n	Semiconductor material with electron majority carriers.
n^+	Heavily doped *n* type semiconductor.
n_I, n_n, n_p	Number of free electrons in intrinsic; *n* type or *p* type semiconductor.

N	Fourth electron energy shell.
N_e, N_h	Number of free electrons or holes per cubic metre.
N_A, N_D	Number of impurity atoms (acceptor or donor).
P	Second sub-shell.
p	Semiconductor material with hole majority carriers.
p^+	Heavily doped p type semiconductor.
P_I, P_n, P_p	Number of free holes in intrinsic; n type or p type semiconductor.
P_c	Power dissipation at the collector (watt).
q	Electric charge (coulomb).
Q	Voltage magnification of a tuned circuit.
ρ	Resistivity (ohm-metre).
r	Distance of electron from nucleus.
r_a	Forward slope resistance of diode (ohm).
r_b	Total base resistance ($=r_{b'}+r_{bb'}$).
$r_{b'}$	Intrinsic base resistance.
$r_{bb'}$	Extrinsic base resistance.
r_c	Collector diode resistance (small signal).
r_e	Emitter diode resistance (small signal).
r_z	Reverse slope resistance of zener diode.
R_B	Ballast resistance.
R_H	Hall coefficient (metre^3/coulomb).
R_L, R'_L	Load resistance, or effective load resistance (through transformer).
R_{OUT}	Output resistance (particularly of stabilizer circuit).
R_{CE}	Total collector to emitter load resistance.
R_s	Series resistance of tunnel diode.
s	First sub-shell.
S	Voltage stability ratio (usually percentage).
t_0	Output pulse length (sec).
τ	Time for complete reversal of flux.
T	Absolute temperature (degrees Kelvin, about -273°C).
T_{amb}	Ambient temperature (°C).
T_j	Junction temperature.
Φ	Flux in transformer core (weber).
v	Linear velocity of electron (metre/sec).
$\hat{v}$	A.C. peak voltage.
v_e, v_h	Drift velocity of electron or hole.
$v_{cb'}$	Alternating voltage from collector to internal base.
$v_{eb'}$	Alternating voltage from emitter to internal base.
V'_c	Collector breakdown voltage.
V_{BE}	Direct base to emitter voltage.
V_{CB}	Direct collector to base voltage.
V_{CC}	Supply voltage.
V_{CE}	Direct collector to emitter voltage.
V_{CS}	Bottoming (or saturation) voltage of transistor.
V_{EB}	Direct emitter to base voltage.
V_s	Forward starting voltage of diode.

V_Z	Reverse breakdown voltage of zener diode.
Y_{11}, Y'_{11}, Y''_{11}	A.C. input admittance, with output short-circuited (common base, emitter or collector).
Y_{12}, Y'_{12}, Y''_{12}	A.C. feedback admittance, with input short-circuited (common base, emitter or collector).
Y_{21}, Y'_{21}, Y''_{21}	A.C. forward transfer admittance, with output short-circuited (common base, emitter or collector).
Y_{22}, Y'_{22}, Y''_{22}	A.C. output admittance, with input short-circuited (common base, emitter or collector).
Z_{11}, Z'_{11}, Z''_{11}	A.C. input impedance, with output open-circuited (common base, emitter or collector).
Z_{12}, Z'_{12}, Z''_{12}	A.C. feedback impedance, with input open-circuited (common base, emitter or collector).
Z_{21}, Z'_{21}, Z''_{21}	A.C. forward transfer impedance, with output open-circuited (common base, emitter or collector).
Z_{22}, Z'_{22}, Z''_{22}	A.C. output impedance, with input open-circuited (common base, emitter or collector).
ω	Angular velocity (radian/sec).

Index

SIGNAL FLOW ANALYSIS

J. R. ABRAHAMS
B.Sc.(Eng.), A.C.G.I., A.M.I.E.E., Senior Lecturer, Enfield College of Technology
and
G. P. COVERLEY
B.Sc., M.I.E.E., Research Scientist, Northern Electric Company, Canada

Signal flow graph methods are presented in a simple manner in this book. The terms used in signal flow graphs are defined, and the manner in which these graphs may be simplified to give the required solution is shown. The way in which these graphical methods may be applied to a wide variety of circuit problems in electrical and electronic engineering is explained, and there are sufficient examples to prepare the reader for the analysis of more complex electrical and mechanical systems.

The approach and presentation is somewhat different in that the use of matrix algebra is avoided. This should be helpful to many engineers who are not mathematically inclined. The chapter on transistor circuits includes some of the latest ideas in this field. Up to ten problems are given at the end of each chapter, and there are a number of practical examples in the text. A numerical or graphical answer to each problem is given in an appendix. The book is intended for the engineering student who is in the course of studying for a degree or diploma, or for the practising engineer who wishes to find a speedy solution to an electrical problem.

ELECTRONIC COMPONENTS, TUBES AND TRANSISTORS

G. W. A. DUMMER
M.B.E., M.I.E.E., Sen. Mem. I.E.E.E., M.I.E.R.E.

G. W. A. Dummer writes in his preface to this book, "Component parts are the building blocks of all electronic equipment. This book is intended to bridge the gap between basic measurement theory of resistance, capacitance and inductance, and the practical application of electronic components in equipments." Whilst the basic theory of electron tubes and semiconductors is available elsewhere, this book emphasizes the more practical aspects.

The reliability of complex electronic equipments depends a great deal on the choice and correct use of the individual component parts, incorrect use often leading to expensive repairs. The essential characteristics of each main type of component, tube and transistor are therefore summarized and references are given for further reading if desired.

The book is written as an undergraduate and technical college textbook, and will help the student over the difficult problem of choosing the right component and device for his particular application.

Printed in Dunstable, United Kingdom

91338431R00181